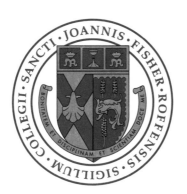

THE MISUNDERSTOOD GENE

THE MISUNDERSTOOD GENE

Michel Morange

TRANSLATED BY MATTHEW COBB

HARVARD UNIVERSITY PRESS

CAMBRIDGE, MASSACHUSETTS   LONDON, ENGLAND   2001

Originally published as *La Part des Gènes,*
© Éditions Odile Jacob, 1998

Library of Congress Cataloging-in-Publication Data

Morange, Michel.
 [Part des gènes. eng]
 The misunderstood gene / Michel Morange ; translated by Matthew Cobb.
  p. cm.
 ISBN 0-674-00336-5
 1. Genes—Popular works. 2. Molecular genetics—Popular works. I. Title

QH447 .M6713 2001
572.8′6—dc21

                                                    00-053916

CONTENTS

Preface

My aim in this book is to put forward a new vision of genes and their function based on recent results from molecular biology. Two things made me write the book. Firstly, the concept of the gene that is used both by the general public and by many scientists is completely outmoded. It is as though one of the key findings of the early years of molecular biology—the discovery at the beginning of the 1940s that the sole function of a gene is to enable a protein to be synthesized—simply had not taken place. And yet it is impossible to discuss the role of genes in any biological process without referring to the protein, the direct product of the gene, which participates in the realization of that process. As a result of this apparent ignorance, discussions of the role and importance of genes, whatever the author's standpoint, are often as relevant to the real world as debates about how many angels can dance on the head of a pin.

My second motivation is simpler: the recent findings from the new technique of genetic "knockouts" are largely unknown outside of the small circle of readers of scientific journals. This is understandable, given that the technique has been used for only a decade. But these results are of such importance for our understanding of gene function that they are crying out to be presented to the general public.

I would like to thank everyone who helped me think through the ideas that are presented here, and first and foremost Peter Beurton, whose multidisciplinary research on the history and philosophy of the concept of the gene greatly influenced my thinking. However, this book has another, more scientific, objective: to show how recent findings from biology oblige us to change our view of the role of genes in complex processes. In this respect, it bears the indelible yet ineffable mark of my many discussions with Charles Galperin and Nadine Peyriéras on the concept of the developmental gene.

I would also like to thank all my students, in particular those in the History of Epistemology course, who not only heard my earliest views on these issues but also helped to develop them. Consciously or unconsciously, I have used many of their questions and ideas in the pages that follow.

Two years running, the audience at the Seminar in the History and Philosophy of Science at the Ecole Normale Supérieure heard various parts of this book presented to them. Anne Fagot-Largeault, Bertrand Saint-Sernin, and Daniel Andler helped me to develop the clarity and precision philosophers require. Anne Fagot-Largeault's remarks on the first draft of the manuscript were particularly useful.

I am also grateful to all my scientific colleagues who, in one way or another, helped me in my thinking: the teachers in the Developmental Biology departments of the Paris VI and Paris XI Universities; Alain Prochiantz, who gave the book its French title; André Klarsfeld and François Schächter, who helped me understand the genes of rhythm and of aging, respectively. Ruth Scheps, by inviting me to write an article, helped me to deepen my knowledge of the genes involved in programmed cell death. And I trust those whom I have forgotten to thank will not hold it against me.

Finally, I would like to thank Gérard Jorland, who supported this project from its earliest stages and who has substantially contributed to the final form, and Matthew Cobb who has not only translated the French version into clear English but has also enriched it by his pertinent comments. The English edition is substantially different from the original version published in French. There are two reasons for this: first, more than two years have gone by between the two publication dates, and in these scientific domains that is a long time indeed. Second, I have—I hope—improved the text thanks to comments by Matthew Cobb, Michael Fisher of Harvard University Press, and referees who read the initial translation.

THE MISUNDERSTOOD GENE

What a wonderful thing it is that the drop of seed from which we are produced should carry in itself the impression not only of the bodily form, but even of the thoughts and inclinations of our fathers! Where can that drop of fluid matter contain that infinite number of forms? . . . 'Tis to be believed that I derive this infirmity from my father, for he died wonderfully tormented with a great stone in his bladder . . . I was born above five and twenty years before his disease seized him, and in the time of his most flourishing and healthful state of body, his third child in order of birth, where could his propension to this malady lie lurking all that while? And he being then so far from the infirmity, how could that small part of his substance wherewith he made me, carry away so great an impression for its share?

> *Michel de Montaigne, "Essays" (1575),*
> *translated by Charles Cotton*

Whatever opinion one may have concerning the nature of the soul, whatever skepticism one may have adopted, it would be difficult to deny the existence of intermediate intellectual organs, necessary even for the thoughts apparently most removed from things of the senses. Among those who have devoted themselves to profound speculations, there is none to whom the existence of these organs has not been manifested, often by the fatigue they feel.

> *Condorcet, "On Public Instruction" (1791),*
> *translated by Keith Michael Baker*

INTRODUCTION

The two quotations on the preceding page embody one of the key questions of modern biology and provide the framework for this book. A modern geneticist could answer Montaigne's questions. We now know that hereditary transmission takes place through the transfer down the generations of macromolecules called genes, and that semen and eggs contribute equally to this transmission. Genes not only participate in the determination of our similarities and our differences but also account for some of the problems we suffer in old age. Today, we see genes as the one thing that enables life to exist, in all its forms—even in realms that might be thought to be far from the material world, such as literary or artistic creation. The reason is simple: genes permit the construction of Condorcet's "intellectual organs."

I want to show how recent biological research enables us to study the influence of genes in processes as specifically human as language or intelligence. To get there, we will have to take a journey through genes, their functions and malfunctions, and the biochemical complexity of the cell. This voyage will be as simple as possible, but it cannot be avoided if we are to demystify the function of genes. This book is not a theoretical or epistemological analysis of "the gene,"[1] but above all a study in biology. Unlike many scientific books that deal with similar subjects, I will not dwell on the role of genes in the expression of individual differences.[2] Instead, I will focus on the role of genes in the construction of organisms. During the first half of the twentieth century, classical genetics revealed the role of genes in the control of differences. Molecular genetics subsequently showed the importance of genes in the control of what is common to all life forms. This is the aspect of the role of genes that particularly interests me and is the subject of this book. It is also the aspect that is the least widely understood.

Before going any further, I will describe my conception of the role of genes. The opponents of molecular biology and of genetics claim that there are two types of biologist: those who want to explain the complexity of life simply by understanding genes, and those who situate this complexity at the level of other components of the organism and, above all, in its organization. This dichotomy is absurd and gives a wrong impression of the molecular vision of life that has developed since the 1950s. If molecular biologists had to designate one category of macromolecules as being essential for life, it would be proteins and their multiple functions, not DNA and genes. Genes are important only because they contain enough information to enable the synthesis of these proteins at an appropriate time and place.[3] Proteins carry out their functions only when they are integrated into the hierarchical structure of life— macromolecular complexes, organelles, cells, organs, organisms. All these levels of organization are based on the properties of proteins, but provide these proteins with a framework within which they can carry out their functions.

As we will see, today's genes were not the first step in the history of life.[4] But they have the remarkable property of giving each organism the ability to transmit the information necessary for its offspring to make the same proteins and to adapt the rate at which they are synthesized in different cells to the needs of the organism. Asking what genes do simply means trying to find out how proteins—the structures of which are transmitted from generation to generation—enable the organism to carry out complex functions. If evolution has given genes their current form, it is because this form provides a remarkably simple and efficient way of reproducing the structure of proteins and the conditions for their expression. On the other hand, this invention has provided biologists with a simple way of studying proteins and the role of these molecules in the functioning of organisms. Thus, genes have pride of place in modern biology not out of principle but out of experimental practice, in terms of both the study of organisms and their possible modification.

The two epigraphs above highlight the different approaches I will use in this book to determine the role of genes in the development and functioning of organisms. The first approach is historical. The quotation from Montaigne reminds us that the questions posed here are not new, even though they have been transformed by the rise of genetics, molecular biology, and now genomics and postgenomics. The questions that surrounded the birth of genetics at the beginning of the twentieth century are still pertinent. The development of the concept of the gene was a slow and difficult process. The debates that took place a century ago, both between geneticists with different viewpoints as to the nature and function of genes and between advocates and opponents of genetics, were not so very different from those that exist today.

The other important aspect of the problem of gene function is experimental and thoroughly contemporary. It flows from the accumulation of experimental results obtained through the use of genetic engineering. Now that whole genomes have been sequenced, the problem of gene function is posed in a very different way. It is increasingly difficult to hide behind the properties of genes that have yet to be discovered in order to explain the functioning of organisms. Already the entire genomes of some bacteria, yeast, a plant, the nematode worm, the fruit fly, and now humans have been sequenced. There is no alternative: the way to understand how organisms function is through their known genes.

However, mere sequencing data will not tell us how organisms function. Instead, I will focus on the rich data that have been obtained from a vast number of studies of natural genetic mutations in both animals and humans, and on the targeted inactivation of genes—"knockout" experiments. These are the experiments that we will learn to interpret in the pages that follow. The harvest of results they have produced is all the more informative in that these findings were often completely unexpected and sometimes even contradicted existing models.

Two concepts of the gene—"variations in genes explain the observed differences between members of a given species" (classical genetics) and "genes make it possible to synthesize proteins" (molecular biology)—have long co-existed in isolation from one another. But recent progress in gene mapping and gene knockouts has changed that. Today, the direct confrontation of these two concepts means that we must try to explain the role of genes in complex processes in terms of their simplest functions.

My description of gene function is therefore not abstract, like that of some geneticists—"genes constitute the space of possibilities that is available to the evolution of life" or whatever—but as concrete as possible, giving a precise image of their functions in the most fundamental life processes: development, aging, learning, behavior, the establishment of biological rhythms, and so on. In fact, an abstract view of gene function is meaningless, for the simple reason that the characteristics of organisms are intimately linked to the nature of the molecules that constitute them. The genome "speaks biochemistry, not phenotype."[5] To use an analogy from computing, if we consider that organisms are machines that process and transmit information, the material that makes up life is at least as important as its program, if not more so. Organisms are algorithms that are incarnated in DNA molecules and in proteins.[6] Without such an incarnation, organisms would not exist. There is far more richness and meaning in the structure of a protein or in a single knockout experiment than in all the theoretical speculations on the role of genes. The real world is far more meaningful than the possible worlds that are dreamt of in our philosophy.

It would be wrong to try to avoid the fundamental question "What do genes do?" In Chapters 1–3, after a brief description of the discovery of the chemical nature of genes and their role in coding for proteins, I describe the various ways that people try to dodge the question of the role of genes—for example, by saying that because the concept of the gene is ambiguous, it has become useless. Recent studies have allegedly reduced the importance of the gene, in favor either of the genome or of cellular structures. But as we will see, even if it is impossible to give a general definition of the gene, the concept of the gene, even if it is vague, is indispensable for modern biology. The phenomena of nongenetic heredity are described not in order to deny their existence but to show their limits. Invoking such effects is yet another way of avoiding the problem of the role of genes and the menacing notion of genetic determinism.

Chapters 4 and 5 show how the analysis of genes involved in various pathologies or in targeted genetic inactivation experiments have made our vision of gene function even more complex. I then examine (chapters 6–8) the role of genes in such complex functions as longevity and behavior. In the process, our story passes from situations where the role of genes is quite clear to those where it is more problematic—and more worrying.

The book closes with a discussion of genes and the human race from the point of view of the history of humankind, the limits that genes apparently impose on our freedom, and the dangerous fantasy that has continually reappeared throughout human history: eugenics (chapters 9 and 10). This ideology will be confronted with the models of gene action that are described throughout the book.

Is it reasonable to devote so much attention to recent discoveries about genes and gene function? Many studies in the social history of science have questioned the special status of scientific knowledge—often with a degree of success. From this point of view, scientific theories and models make up only a part of all possible theories and models—those that history has retained for contingent reasons, not because of any inherent value. The interpretation of the functions and evolution of life from a genetic point of view would thus be merely another theory, among many possible such theories, put forward for a variety of reasons.

However, even if scientific theories are only one reflection of reality, it would be illusory to think that by replacing the genetic description of life with another description, we would escape from the threat of genetic determinism. Fleeing into a relativist interpretation of scientific knowledge may seem to be an easy way out, but in fact it is no way out at all. The threats will still exist, the new technologies of gene manipulation will still be there, ready to be wielded for better or for worse.

I think that the greatest resistance to a frank study of the role of genes comes from the fear—conscious or unconscious—of the answers we might find, and the possibility that they could threaten our individual liberty and the foundations of democracy. Differences among genes might create differences among people. However, the greatest danger in fact would be not to address squarely the question of the role of genes at a time when the CIBA Foundation organizes a meeting on "The Genetics of Criminal and Antisocial Behavior"[7] or when every week the leading scientific journals—*Nature, Science,* and others—publish articles that describe the localization of genes that allegedly control human personality traits or are involved in the genetic control of behavior. Some researchers have even claimed that happiness depends on our genes and not on positive life events.[8]

My aim is to face the problem of genetic determinism squarely and, as Steven Rose has put it, to go "beyond" it.[9] True, we are not in our genes.[10] Despite what certain behavior geneticists seem to believe, we are not "ready-made from the factory."[11] But we are not without our genes, either. We are going to have to make do with them—for the better.

CHAPTER ONE

THE CONCEPT OF THE GENE

The history books tell us that the laws of heredity were discovered in 1865 by a Moravian monk named Gregor Mendel, who conducted experiments with peas. Mendel showed that:

- an organism's "characters"—their color and form, in the case of peas—are due to the action of a pair of determining factors (what we call genes)
- one of these factors comes from the male parent, the other from the female parent
- these factors do not mix but are transmitted intact and in a random fashion.

Mendel's results were ignored and then rediscovered in 1900 by Carl Correns, Erich von Tschermak, and Hugo de Vries. But genetics really took off only at the end of the first decade of the twentieth century, when Thomas H. Morgan decided to study the organism that would eventually be identified with the discipline, the tiny fruit fly *Drosophila.* In the space of a few months Morgan and his team showed that genes were carried on chromosomes, and then they drew up the first genetic maps.[1]

DEVELOPMENT AND HEREDITY

Why did it take so long for the laws of heredity to be discovered? The fact that children resemble their parents has been known as long as humanity has existed. However, there are many cases where children do not look like their parents or, more curiously, where they resemble their grandparents, uncles, or aunts. All the models that tried to explain the phenomena of heredity were thus handicapped by both the known facts and their complexity. They were also hampered by a series of preconceptions: the inheritance of acquired characters—that is, the ability of organisms to pass along to the next generation any behaviors or other characteristics they acquire during their own lifetime, such as a specific physical aptitude or a gift for music or painting, was considered self-evident in the eighteenth and nineteenth centuries. Even Darwin did not question it.

But the most difficult task that had to be accomplished before the science of heredity could be established was separating the study of the transmission of characters over the generations from the study of development. This division was logically absurd, but it was historically and scientifically necessary.

Two successive theories had been proposed to explain development. The first theory—preformationism—proposed that, from the very beginning, the adult was already totally formed in the shape of "animalcules" present in the paternal semen or in the maternal eggs. Embryonic development was merely growth, not transformation. Today this theory appears strange: it supposes that the gonads of the developing organism contains the members of future generations, one within another, like an infinite series of Russian dolls. However, at the time this seemed much less absurd than it does today, when we know that the organization of matter into atoms places an absolute barrier to such *emboîtement,* as it was called. The theory's long survival and the support it received from many enlightened thinkers can be explained by the fact that the only rival theory, epigenesis, also raised many difficulties. Although epigenesis would eventually replace preformationism, following detailed studies of development, this initially appeared unlikely because epigenesis explained development in terms of the progressive formation of the organism and was thus unable to explain resemblance. If the organism is built up bit by bit, going through a series of different stages that generally are very different from the final form, how can it (sometimes) so closely resemble one or other of its parents for one or more characters? At what point could resemblance have been introduced?

The birth of a science of heredity meant that this question could be side-stepped—provisionally, at least. The sole aim of the new discipline was the study of similarity, without worrying about the mechanisms involved in creating that similarity. Scientific progress is often, if not always, a result of setting certain problems to one side. "Finding out" involves choosing what exactly, among the complexity of reality, one wants to know and accepting that one will not understand the rest, which will thus be neglected or even forgotten. This choice, which is more unconscious than conscious, more social than individual, can often be understood only by studying the historical context.

The beginning of the twentieth century was more favorable to the birth of an autonomous science of heredity than was Mendel's time. Interest in the transmission of characters from parents to children had grown. Writers and artists often highlight and even anticipate the preoccupations of their contemporaries. For example, Emile Zola reflected the wide interest in heredity through his description of the Rougon-Macquart family in his cycle of novels. This interest was colored with anxiety, because many people thought that among humans, faults were transmitted over the generations more often than gifts, handicaps more often than advantages, and that, bit by bit, this would lead to the degeneration of the race. This would occur because natural selection, which Darwin had shown played an essential role in the evolution of species, was prevented from fully acting in humans by cultural and social practices, which opposed, for instance, the elimination at birth of babies affected with serious physical defects. In 1883 Darwin's cousin, Francis Galton, took up this challenge and founded the science of eugenics, both as a theoretical investigation of heredity and as a practical discipline with the objective of improving hereditary transmission in humankind.[2]

Mendel—a botanist—had crossed pea plants in order to study the stability of the resultant hybrid forms and to better control the production of new plant varieties. His successors were interested in the transmission of hereditary variations over the generations as species, and in particular the human species, evolved. Despite the legend, Mendel was thus neither unknown or forgotten—his "rediscoverers" knew his work. Mendel had just simply been ignored, because in the second half of the nineteenth century his work did not excite the same interest as it did among scientists at the beginning of the twentieth century.

GENETICS TAKES ITS FIRST STEPS

Genetics did not appear overnight. Mendel's genetics—that of his rediscoverers—was not the same as the genetics that school children learn today. The term "genetics" was introduced in 1903, by William Bateson, and the word "gene" was not adopted until 1909. The early articles and books by geneticists deal only with the transmission of characters over the generations. At its birth, genetics was a science of similarities, not—as it would later become—a science of the mechanisms that produce these similarities.

The introduction of the term "gene" was a response to a difficulty encountered by the early geneticists: the transmission of characters is accompanied by variability. The Danish geneticist Wilhelm Johannsen was the first to show that this variability had two origins. There is a variability linked to the environment, to everything that surrounds the character. And there is also a hereditary variation due to the transmission of factors that are slightly different. The distinction between characters and the factors that control their formation (baptized "genes") was the product of a series of observations on the transmission of variability in plants, rather than the logical requirement of defining that which is transmitted from generation to generation and which ensures the continuity of characters.

T. H. Morgan, who was to be the founder of the most productive school of genetics, was initially highly skeptical of genetics,[3] and his doubts are extremely important for understanding how researchers viewed the new science. Before 1910, Morgan had two key criticisms. The first—which he stated explicitly—was that geneticists claimed to have the answer to everything. Faced with a biological phenomenon that required explanation, they had a ready-made answer—genes. Indeed, it is easy—too easy—to explain an effect by appealing to some "hidden" causal factor. And if one factor was not sufficient to explain the observed phenomenon, then a second could be invoked, and then a third, and so on. If Karl Popper had written his *Logic of Scientific Discovery* at the time, Morgan would undoubtedly have said that the burgeoning science of genetics was unfalsifiable.

Less explicitly, but more fundamentally, Morgan criticized genetics because it was a kind of throwback to preformationist theories, in that it sought to explain the characters of adult organisms simply by invoking the transmission of factors—genes—as if these adult characters were already somehow contained in the genes. These criticisms are remarkably similar to those aimed against genetics today. For instance, the opponents of behavior genetics criticize the attempts of certain practitioners of this discipline to explain every aspect of human behavior by the existence of specific genes, the nature of which remains unknown but which can nevertheless miraculously encode a complex behavior.

However, Morgan was convinced by the results of his initial experiments on fruit flies. The transmission of eye color or wing shape could easily be explained if the factors—genes—responsible for these characters were associated with the chromosomes present in the cell nucleus. The fact that one of these chromosomes existed in two different forms in males and females could also explain why a given individual was of a given sex.

Morgan soon noticed that in a small proportion of cases, characters on the same chromosome could be inherited independently. Part of his genius was that he quickly found an explanation for this phenomenon: it is the result of "crossing-over"—the mutual exchange of parts of the two chromosomes in a given chromosome pair. This cellular event had recently been discovered and could be observed during the formation of the sex cells. This explanation made it possible to link the recombination rate to the physical distance between genes on the chromosome and thus to draw up the first chromosome maps.

This incredibly productive period in which the principles

and main techniques of genetics were developed can be precisely situated between 1908 when Morgan began to study *Drosophila* and 1915 when he published *The Mechanism of Mendelian Heredity* together with his three brilliant students Calvin Bridges, Alfred Sturtevant, and Hermann Muller.[4]

THE REIFICATION OF GENES

Thanks to Morgan, genetics began the long process that was to lead slowly to the transformation of the gene into an object, followed by the chemical and functional description of this object—a process we might call reification. At first it was not at all clear that the gene was in fact a material object. Perhaps it was something else, such as a vibration or a form of organization. Scientists such as William Bateson and William Castle tried to slow down this reification of the gene by raising doubts as to the stability of genes and their chromosomal localization. But these doubts faded in the face of overwhelming evidence provided by Morgan's group, in particular cytological data that clearly demonstrated the relation of genes and chromosomes.[5]

The criticisms of Richard Goldschmidt, on the other hand, were both broader and more interesting. He attacked the notion of the gene as a particle of heredity, emphasizing instead a dynamic and physiological vision of gene action.[6] And even more interesting is the lack of enthusiasm for genes as material objects that was shown by the most orthodox geneticists. Edward East, for example, believed that a genetic interpretation of the hereditary transmission of quantitative traits was a mere hypothesis, a practical way to explain observations that in no sense required genes to actually exist as physical entities.[7]

More than 25 years after beginning his work on *Drosophila*, Morgan, working in this skeptical atmosphere, still thought it was possible to practice genetics without believing in genes. This prudent position was no doubt partly due to the historical context—in particular seeking to avoid the errors of preformationism. But above all it reflected the difficulties that the reification of genes would encounter when it came to finding out what they really did.

A few key discoveries that were responsible for the reification of genes will show that this process was not the result of a scientific strategy but rather of a series of unexpected findings. Against all their expectations, geneticists came to be convinced that genes were real, and then that they were made out of DNA. In 1927 H. Muller showed that X-rays had mutagenic effects and could thus transform genes. This was important because it provided an easy way of creating large numbers of new mutants. Above all, it showed that the gene was indeed a physical entity, made of a material that was sensitive to X-rays. Just as X-rays had enabled physicists to open the doors of the atomic nucleus, Muller's discovery made it possible to imagine opening the black box of the gene.

Second, the discovery of giant chromosomes in the salivary glands of *Drosophila* made it possible to study the fine structure of chromosomes directly, under a light microscope. At the beginning of the 1930s, T. S. Painter observed "chromosome inversions" under the microscope, showing that chromosomes could break up and then be repaired such that one or more stretches of genetic material were no longer in the "correct" orientation, thus preventing crossing-over between the two chromosomal forms (inverted and "normal"). The existence of such a process had been invoked theoretically by geneticists since the beginning of the studies on *Drosophila* in order to explain some of their results. Painter provided the visual proof of its existence.

Having been identified as a target, and then definitively localized to the chromosomes, the gene became the focus of attention for molecular biologists, as this group of scientists would later be called. The aim of this new kind of biologist was to explain the fundamental phenomena of life through the properties of its constituent macromolecules. In characterizing the constituent elements of organisms, biochemists had already shown that proteins were extremely important, as both the material that makes up the cell and the catalysts that give the cell its extraordinary powers of chemical synthesis. In the 1920s, when biologists first tried to describe the gene's chemical nature and role, they quite naturally suggested that genes were proteins. Up until the 1940s, this protein or enzymatic theory of the gene was dominant, and it was reinforced by experiments carried out in 1940 on a large number of mutations in a fungus, *Neurospora crassa,* which showed that there was indeed some kind of close relation between genes and enzymes.

The protein theory began to fall out of favor, however, when the American microbiologist Oswald Avery showed that it was possible to transform the hereditary properties of a bacterium, the pneumococcus, by adding a pure DNA molecule to the culture. DNA gradually replaced protein as the candidate molecule constituting genes. In 1953 James Watson and Francis Crick proposed that DNA has the form of a double helix—a structure which made it possible, although not without some difficulty, to explain the self-replicatory properties of genes.[8] The only possible function for a molecule that was so long, with a regular but nonmonotonous structure, was to encode information. George Gamow, a physicist, was the first to put forward the hypothesis of a genetic code.[9] He suggested that there was a relation between the sequence of base components—the nucleotides—along a DNA molecule and the sequence of amino acids that form proteins. The code theory was quickly taken up and developed by Francis Crick and was experimentally confirmed eight years later. The genetic sequence of four bases—A, C, G, and T—is made up of "words" consisting of three bases. Each "word" corresponds to an amino acid, and the sequence of "words" thus makes up the protein coded for by the gene. With a few extremely minor exceptions, this genetic code is common to all known life forms.

With the deciphering of the genetic code, genes were attributed the function that they still have in contemporary biology: they contain information that permits the synthesis of proteins. Each specific protein carries out any one of the many structural or enzymatic functions that this class of macromolecule can fulfil, in all or some of the organism's cells.

POPULATION GENETICS AND NEO-DARWINISM

This succinct survey of the key discoveries that led to the characterization of the fine structure of the gene and its molecular role within the cell has left to one side a whole branch of genetics known as population genetics or evolutionary genetics.[10] This science is not interested in the biochemical nature of genes nor their molecular functions but in their distribution in a given population and in the variations shown by this distribution over the course of evolution.

To understand how the links between genetics and evolutionary theory (Darwinism) were forged, we have to return to the middle of the nineteenth century. Having proposed that evolution was the result of the natural selection of the fittest, Darwin hesitated for some time over the nature and source of the hereditary variations that were the target of natural selection. If these variations were large, it was easy to see how natural selection could "choose" the fittest. However, if this were the case, the ability of organisms to change over evolutionary time would be entirely dependent on these miraculous variations. Darwin, who had tried to remove God from the description of organisms, felt that such macrovariations had an almost supernatural nature. So he opted for the hypothesis that natural selection targeted variations that were extremely small. This hypothesis corresponded more closely to his model of hereditary transmission. However, some elementary mathematics showed that under this model new variations within a population would be smothered by the weight of old characters, and after a few generations even favorable variations that had arisen by chance would disappear from the gene pool.

This very real difficulty tipped the balance once again in favor of large variations. One of the initial motivations of the early geneticists such as de Vries and Morgan had been to discover and follow the hereditary transmission of variations that were sufficiently important to play a role in evolution. Having failed to find such variations, and against his better judgment, Morgan turned to the study of the inheritance of eye color in *Drosophila*. It was to be several years before the geneticists, with the substantial help of statistics, showed that the frequency of the different forms (alleles) of a gene could be modified upward by selection, even if the advantage provided by a given form was very weak. The first studies by Russian population geneticists, cut short by the rise of Stalinism but continued in Germany by Nikolaï Timofeef-Ressovsky and in the United States by Theodosius Dobzhansky, removed population genetics from its usual domain—equation-covered sheets of paper or, at best, the laboratory—and put it into the field. The models that were elaborated at this stage were used to explain the evolution of organisms in the wild, in nature. They were also used to interpret, retrospectively, paleontological data.

Population genetics was enriched and changed by the naturalists' studies, and the synthesis of Darwinian evolutionary theory with Mendelian genetics was advanced. Simple changes in gene frequencies of large populations were replaced by more complex mechanisms, such as the geographic isolation of small populations, which better explained the formation of new species. Instead of reasoning on the basis of the selective advantage of a particular form of a given gene, geneticists learned to take into account the selective advantage coming from the existence of various forms of that gene. They showed that natural selection does not select between different forms of isolated genes but between combinations of different alleles of the totality of genes that constitute an organism. In other words, organisms as a whole, not isolated genes, are the targets of selection.

Over recent years, this neo-Darwinian theory has been repeatedly attacked. Chance has begun to assume an important place in evolutionary theory. The work of biochemists and molecular biologists has revealed the existence of a large number of silent mutations that are not exposed to selection pressures within the environment. Even though natural selection is the key force shaping organisms, the presence of a particular form of gene in a given population is often due more to chance, or rather to history, than to pressure from natural selection. Genomes are full of such silent variations, trapped there rather than selected on the basis of any advantage they might confer on the organisms that carry them. Neutralism, as this concept is called, emphasizes the role of chance in evolution and downplays that of natural selection.[11]

According to other critics, the structural constraints implied by the development of the individual limit the role of natural selection to merely sifting out the acceptable from the unacceptable, thus limiting its adaptive power. The discovery and characterization of fossils from the Cambrian period, when multicellular life exploded onto the planet, have revealed an incredibly rich fauna, with a range of body plans far more diverse than those shown by current life forms. This could imply that this initial diversity was not removed by natural selection owing to some failure at being adaptive but rather was decimated by blind chance.[12]

However, none of these criticisms, all of which tend to minimize the role of natural selection, have fundamentally shaken the neo-Darwinian edifice. The critics have no alternative model to propose. Furthermore, two of the key founders of the synthesis—Dobzhansky and in particular the naturalist Ernst Mayr—did in fact accord chance an important role in evolution. Their theory was sufficiently large—some might say vague—to accommodate what later critics were to consider decisive criticisms.

The robustness of neo-Darwinian theory should not blind us to the fact that it was built upon, and is still based upon, a conception of the gene that pre-dates molecular biology. Like all contemporary biologists, population geneticists accept the results of molecular biology and agree that genes are stretches of a DNA molecule that code for proteins. But this is not the definition of a gene they use when building their models. For population geneticists, a gene is simply a black box that can exist in several stable forms, each of which may give the organism different properties. These forms confer different advantages on the organisms that carry them, in the context of the environment in which the organisms live, and this differential is the substrate on which natural selection acts.

There is no reason why population geneticists should not go beyond the simple definition of genes involved in an evolutionary process and localize them on chromosomes and then define their molecular nature. However, even though these two visions of the gene—that of molecular biologists and that of population geneticists—may sometimes converge, they remain fundamentally different because these two kinds of biologists have divergent views of the importance of the nature of genes and thus of the products they code for. For population geneticists, the nature of genes is entirely secondary. For them, the key property of the gene is not its nucleotide composition but its ability to replicate itself (in the context of the organism that contains it) and its ability to vary. The biochemical composition of the gene and of the protein it codes for is so unimportant to this approach that computer programs can effectively simulate the processes of evolution, by according the gene merely these two properties. The reason why population geneticists are uninterested in the nature of the gene is that the range of possible genetic variation is, for them, if not infinite, at least sufficiently vast for the evolution of organisms to be unlimited.[13]

For molecular biologists, by contrast, decoding genes and understanding the nature of their products is essential. Not because the analysis of their nature has revealed rules or principles that can explain the evolution of life but because the conservation of certain molecular structures and their reuse throughout evolution—sometimes to very different ends—suggests that there are limits to what François Jacob called evolution's "tinkering."[14] Understanding life and its evolution is inextricably linked to understanding the elements that make up organisms. Daniel Dennett has suggested a striking image to describe how evolution proceeds: it continually plays with "Mendel's library" which is present in each organism.[15] For population geneticists, this library is sufficiently large that we do not to need to have read all the books in order to explain how organisms evolve. For a molecular biologist, although the evolution of life is not determined by the information contained in the library, it is nevertheless dependent on that information.

This memorable image illustrates the importance of linguistic and computing metaphors in modern biology and, in particular, in the study of gene function.[16] And it is the very power of the metaphors that are linked to the concept of the gene that strikes fear into the hearts of those who worry about the direction of genetic research.

Computing analogies have accompanied the history of molecular biology since its birth.[17] Computing and molecular biology, the two scientific developments that most marked the second half of the twentieth century, grew to maturity in parallel with each other. Although linguistic and computing metaphors no longer have the positive influence on the theory of molecular biology that they did in the 1950s, they nevertheless still enable biologists to describe their work to nonspecialists.

Two examples will show how these metaphors fascinate biologists and nonbiologists alike and provide a mistaken vision of how organisms function. The first is the parallel that is drawn between the genome—the totality of genes—and a book. This metaphor elevates genes above all other cellular components, which, as we saw in the Introduction, is a misguided way of looking at the internal organization of organ-

isms. It has, however, stimulated interest in the various ge-
nome-sequencing projects. After all, have we not set about to
decode the great "book of life"? Western civilization is the
product of the meeting of two great traditions: the Judeo-
Christian tradition, which puts the Holy Book at the heart of
the relation between God and humanity, and the Greek tradition, according to which natural order is the result of a language or "logos." Given this background, it is hardly surprising
that the "text" metaphor seemed like such a "natural" way to
explain the workings of genetics.

The second omnipresent metaphor is that of the computer
program. As François Jacob pointed out in 1970, "Everything
suggests that the logic of heredity can be taken to be that of a
calculator. Rarely has a model imposed by an epoch found a
more faithful application."[18] The idea of a program seemed to
apply particularly well to the regular process of embryogenesis. However, this computer-program metaphor in fact
has several weaknesses. First, it suggests that development
can be directly decoded from the genes. At best, knowledge of
genes can reveal the structure of the proteins that lead to the
progressive construction of the organism through their complex action in cells, tissues, and organs. The program metaphor also suggests that, in the organic world as in the world of
computing, it should be possible to distinguish the software
from the hardware. But genes and proteins are not software
and hardware. Their functions cannot be separated from the
material of which they are formed. They exist at the same
level of abstraction—they are both macromolecules—and are
in close interaction with each other. The idea of a program in-
troduces a hierarchy between these two elements of the organism: the program (DNA) appears to command proteins,
which take on the role of mere executive subordinates. But

DNA is not the proteins' "superior." In one sense, it is the proteins' slave, in that proteins are required if DNA is to reproduce. In the history of life, DNA did not precede proteins.

If we have to use a metaphor to describe the role of DNA and genes, that of memory is clearly the most appropriate.[19] DNA is the memory that life invented so that, at each generation, its active agents—proteins—could be efficiently reproduced.

Should we rid modern biology of its computing metaphors in order to avoid being misled? Some people think so. However, it is very unlikely that biologists, who have never been much bothered by impure, vague language, will consent to do this.

CHAPTER TWO

CAN WE GET RID OF THE GENE CONCEPT?

In Chapter 1, we saw how molecular biologists first determined the chemical nature of genes and then studied their function, thus completing the reification of the determinants of heredity begun by Mendel nearly a century earlier. At the beginning of the 1960s, the gene was a fragment of a DNA molecule that coded for a protein. Electronic microscopy, pushed to its limits, allowed us to see the molecule's shadow. Its chemical structure and its function were thus perfectly defined.[1]

Strangely enough, the 1960s also marked the beginning of the deconstruction of the concept of the gene.[2] It gradually lost its simple, unequivocal meaning and became much looser.[3] The first step was the demonstration by François Jacob and Jacques Monod that the activity of genes is controlled by regulatory elements situated in the DNA molecule itself, just upstream from the genes. These regulatory elements are necessary for the correct expression of the gene. The boundaries of the gene began to get hazy. This questioning of the nature of the gene might seem of little consequence: surely all that was necessary was to redefine a gene as the totality formed by the coding sequence *and* the regulatory signals? However, as the nature of the regulatory sequences became known, first in bacteria and then in higher organisms, matters became increasingly complicated. The regulatory sequences turned out to be very complex and could be located far away from their target gene. They could even be common to several different genes. The boundaries of the gene were starting to crumble.

These results were obtained with the new techniques of genetic engineering developed in the 1970s. These methods made it possible to isolate genes from higher organisms (including humans), whereupon it was discovered that these genes are "in bits;" that is, they are composed of several separate sections of coding or regulatory portions (called exons), separated by long sequences with no apparent function (introns). But once again, it seemed possible to accommodate this discovery by simply changing the definition of the gene and describing it in terms of the totality of exons and introns.

Unfortunately for this redefinition, it next became apparent that cells can regulate the way in which the sequences in the exons are stuck together. From a given DNA fragment the cell can make several different messenger RNAs and thus several different proteins, each of which can have different functions. Cells have exploited this process of differential splicing of precursor RNAs particularly in the case of genes coding for transcription factors and for components of the extracellular matrix—the external support of secreted proteins which gives tissues their structure.

Further discoveries showed that in some cases the same DNA fragment could produce two completely different proteins by slightly changing the "reading frame"—the starting point of the DNA sequence. This possibility has been exploited by viruses, for which it is an obvious advantage to store the maximum amount of information in the most limited place. How could natural selection "choose" between variants if the same stretch of DNA corresponded to two different functions?

Furthermore, the structure of the genome is not stable

over the life of the organism, and different cells can contain genes with different structures. For example, genomic rearrangements in immune system cells enable them to generate the wide range of antibodies and receptors that are necessary to respond to infection. Other more recently described phenomena, although rare, indicate that the mechanisms which transcribe and translate genetic information, thus producing the correspondence between genes and proteins, are in fact highly complex. For example, a gene may first be copied into an inactive form of messenger RNA which is then slightly modified by the insertion of a number of nucleotides—before being translated into a protein. This "editing" of transcripts was first thought to be restricted to ciliates (unicellular organisms which over a long period of isolated evolution developed very specific mechanisms of replication and functioning). But the same phenomenon has since been discovered in higher organisms, where it controls, for instance, the structure of some brain neurotransmitter receptors.[4]

A number of lessons can be drawn from this list of weird and wonderful examples, which is no doubt far from complete. The first, fairly obvious, conclusion is that there is no universally valid definition of a gene. While one can put forward several different definitions, none of them will include all known cases. Over recent years, the gene has thus become a vague and slippery concept.

Is this good or bad? Some people have rejoiced: if it is no longer possible to define the gene, then there can be no genes for intelligence or homosexuality and the threat of genetic determinism that hangs over humanity would disappear. That would be jumping to conclusions. Common sense—which is as valid in science as elsewhere—suggests that exceptions are less important than the general rule. The concept of the gene, as it was elaborated by molecular biologists in the 1950s, remains valid for the vast majority of cases. And even where it no longer applies, it is still possible to propose local, contextual definitions of a gene, valid for each particular case. Although the implosion of the concept of the gene may be sad from a fundamentalist standpoint, demonstrating that our concepts are never perfect descriptions of the world, it in no way prevents biologists from working with and using the concept of the gene in constructive ways. Even deconstructed and broken, genes remain the favorite toy of molecular biologists.

The fact is that scientists have always been much more comfortable with vague concepts than have philosophers. It has even been argued that this very plasticity of scientific concepts is necessary for scientific work to go forward. Codifying scientific concepts renders them rigid and hampers the permanent reorganization of knowledge that is the motor of scientific discovery. A vague concept is often rich in explanatory potential.

And even were the concept of the gene to disappear, genetic determinism would not disappear with it. The idea that human beings differ by their abilities and that these differences are transmitted to offspring predates genetics. Eugenics was founded by Francis Galton at a time when Mendel's results were still largely unknown. The risks of eugenics would not disappear with the disappearance of genetics.

Genes and Genomes

Faced with the inability of biologists to define exactly what a gene is, some philosophers of science have proposed the following solution: abandon the concept of the gene and simply speak about genomes and bits of genomes.[5] The immediate advantage of such a change is obvious: whereas it is impossible to define a gene, it is very easy to define a genome—it is the totality of DNA molecules transmitted from generation to generation. Encompassing all the genetic material contained in a cell, the genome is clearly and absolutely circumscribed.

The second advantage of a shift from gene to genome is that the genome provides a better starting point for explaining the molecular events that take place during evolution. Whereas such changes often involve limited transformations —the mutation of genes—other events can also take place, such as duplications of genes or of substantial parts of the genome, the juxtaposition of hitherto separate DNA sequences, or the transformation of noncoding sequences into coding ones (the opposite change is also possible). These kinds of transformation are extragenic but intragenomic. Describing evolution in terms of the modification of the genome rather than of genes would thus be more in keeping with the reality of events that take place over the course of evolution.

The third advantage of focusing research on the genome rather than on the genes that make it up flows from the fact that the genome might be more than the sum of its genes. The ordered juxtaposition of genes and the spatial organization that results could have an influence on the activity and properties of genes. This idea is almost as old as genetics itself. Very early on, geneticists discovered what they called "position effects": the activity of a given gene can depend on its position in the genome, that is, on its genetic environment. Richard Goldschmidt took this view as his starting point for his criticism of the gene as particle, arguing instead for a concept of integrated chromosomal function.[6] Genes were found to associate with proteins to form a nuclear macromolecular structure called chromatin that exists in several different forms, some of which completely repress the activity of the genes contained in them. The position of a gene with respect to these different forms of chromatin partially explains the position effects described by geneticists. The structural organization of chromatin is supragenic but intragenomic.

Because of these findings, at the end of the 1970s many biologists felt that the position of a gene in the genome had a more fundamental influence on the gene's activity than its immediately surrounding regulatory sequences. The success of the first transgenic experiments thus came as a complete surprise: in most cases, if a gene was inserted somewhere—anywhere—into the genome, it had the same effect as in its endogenous position. Sometimes its level of activity (expression) was affected by its position in the genome and the controlling sequences which may be present upstream or downstream, but in most cases expression was "normal." These results have since been confirmed over and over again. Without denying the roles of chromatin and of genome organization, their importance is nonetheless limited.

So far, complete genome sequencing of the chromosomes of bacteria, yeast, a plant, the nematode worm, fruit fly, and human has not revealed any supragenic organization, either. Some genes do indeed tend to group together in a given genome, and they often represent the trace of ancient gene duplications or even "horizontal" (interspecific) genetic transfers. Furthermore, in the case of the nematode, the most highly conserved genes tend to be those located in the middle of the chromosomes, where the recombination rate is lowest, whereas the least conserved genes are found on the chromosome arms.[7] Nevertheless, the observations made by the pioneers of genetics remain valid: genes are apparently distributed more or less randomly along chromosomes. Perhaps there are some surprises in store for us with the sequencing of higher organisms, but as far as we know now, the genome is in fact the sum of its genes.

The second argument that appeared to encourage a change of focus from genes to genomes—that many evolutionary changes occur at the level of the genome rather than the gene—is equally weak. Although there are cases where fragments of genes can associate and form new genes, or where noncoding sequences (introns) have been recruited to form coding sequences, such examples are relatively rare. These kinds of mechanisms were probably essential for the evolution of early forms of life, but since then their importance has lessened and today is limited to a few protein assemblages such as the components of the extracellular matrix to which the cells of multicellular organisms are attached. The mass of noncoding sequences that are contained in a cell of a higher organism (more than 90 percent of the DNA in humans) do not seem either to structure the genome or to be a kind of "genetic reserve" used by evolution in order to provide innovation. They are just there. How else can we explain the fact that the proportion of noncoding DNA can show such large variations between organisms with analogous forms of function and development? Indeed, biologists can exploit this variability: Sydney Brenner has proposed that geneticists should pay particular attention to the Fugu, a fish that is apparently as complex as any other "average" vertebrate but which has a genome that is only an eighth of the expected size.[8]

If this large proportion of noncoding DNA is so unimportant, would it not be more sensible for the major genome sequencing programs to concentrate on sequencing the genes and simply skip over the intergenic sequences? In most cases researchers are extremely prudent and prefer to obtain the total sequence, so that any potential role of noncoding sequences can thus be studied subsequently. In fact, this problem is particularly pressing given that the regions between genes are the most difficult to sequence because they are full of repeated bits of DNA. It was in fact argued that the publication of a complete genomic sequence should be suspended simply because a few regions, which are thought to have no function, resisted the efforts of the sequencers. After much debate, scientists working on the nematode genome finally decided to publish a "complete" genomic sequence that nevertheless contains a number of "holes" that make up around 3 percent of the total sequence.[9] The same attitude was adopted for human chromosomes. And the fruit fly genome was released with one thousand gaps![10]

Putting "genome" in the place of "genes" simply because the former can be defined and the latter cannot would therefore resolve one difficulty but at the same time create many more. The following chapters show the usefulness of the concept of the gene both for understanding the genetic bases of a large number of illnesses or for describing development. Replacing the concept of the gene with that of a "fragment of the genome" would not have any particular advantage whatsoever. The only solution seems to be to continue to make use of a concept that is currently impossible to define.

Although the sequencing projects that are decoding the genomes of various organisms have not revealed the existence of any supragenic organization, they have nonetheless provided important results. The first is to give us an idea of the proportion of genes involved in each of the key biological functions. For example, the yeast genome was sequenced in 1996. In such a simple organism—which is nevertheless fundamental in the preparation of many kinds of food—it might be expected that most genes would be involved in metabolism (the transformation of nutritional sources). In fact, such genes represent less than a quarter of the yeast's genes, whereas more than 40 percent of them are involved in the structural organization of the cell and the exchange of signals and macromolecules across the various intracellular compartments.[11] Even a cell as simple as yeast is not simply a bag of enzymes but a highly organized structure. This example shows that life is characterized by organization. Organization allows elementary components—proteins—to participate in complex functions.

A second surprise in yeast was that the total number of genes was higher than predicted from genetic studies. Some of the genes were previously invisible, because when they were deleted in "knockout" experiments (see Chapter 3), there was no detectable effect on the organism. As we will see, this can be explained by the concept of redundancy (different genes may have, at least partially, identical functions) or by the fact that in standardized laboratory conditions a given organism will not express all its genes, as would be the case in the wild.

Initial results from the studies of the genomes of complex higher organisms such as mice or humans have shown that these genomes are the result of repeated partial or total duplications of simpler genomes.[12] The yeast genome is itself the result of an ancient duplication.[13] The most important difference between the genome of a fruit fly and that of a human is therefore not that the human has new genes but that where the fly only has one gene, our species has multigene families.

The most fundamental result of the complete sequencing of simple genomes—bacteria, yeast, nematode, fruit fly—is no doubt that it has shown us not to expect revolutionary results from the study of genomes. Of course, as we have seen, these studies have revealed the existence of a number of hitherto unknown genes, the functions of which are equally unknown, and many of which bear no similarity to any previously described genes. Researchers have underlined the importance of finding these new genes, which were discovered only by a systematic approach. Consciously or unconsciously, this way of looking at things emphasizes the utility of sequencing projects. However, such discoveries are less original than may first appear, and indeed they were predicted.[14] Furthermore, the ratio of known to unknown genes can be seen from two points of view. The glass can be half empty or half full. In fact, the number of completely unknown genes revealed by sequencing is not that large, certainly not larger than the number of genes that are well understood.

More importantly, all these results show that the richness of structure and function of organisms should not be sought in a genetic never-never-land. It is already here, to hand, ready to be read. To understand the language of life, we simply need to understand the meaning and use of words that are already widely known, rather than seeking to discover new words. The most difficult will be to understand how these words are assembled to generate sentences—or, in a nonmetaphoric way, how the simple elementary functions of the genes are organized to generate the complex functions of organisms. We will come back on this problem in Chapter 9, after a long journey through these elementary functions.

Nongenetic Heredity

Before dealing in detail with the question of the role of genes, one last illusion needs to be dispelled. That is the possibility that there might be other forms of heredity than genetic heredity: genes might merely be one of the actors—and not necessarily the main one—involved in hereditary phenomena. Alongside the "sequential" heredity of genes, there could be a structural heredity, a heredity of form.[15]

Against the idea that genes contain all the information necessary for the reproduction of organisms, it is often said that genes alone are inert and that they have never been seen to be active outside of a living cell. This argument is correct, but it does not contradict the idea that genes are the complete and unique memory of life, the one "thing" which, from generation to generation, makes it possible for living organisms to be reconstructed. It does, however, expose the illusion that, taken on its own, memory can be anything other than inert. Genes do not make it possible to create an organism *ex nihilo;* they simply make it possible to transmit an organism's characteristics to the next generation.

The idea that genes are only one component of the hereditary process is as old as genetics itself. The search for another form of heredity, independent of the chromosomes, preoccupied many biologists, especially between the two world wars.[16] This continued even after the pioneer work of molecular biologists, because many scientists found it difficult to accept that the highly organized structures typical of life could be the product of a one-dimensional message, that such forms were not born of another form that served as a matrix. The intervention of another form of hereditary memory apart from the genetic has thus been sought at all levels of organic organization, from the whole organism through the cells to macromolecules.

The most determined advocates of extrachromosomal heredity were the embryologists. In many species, the first stages of development appear to follow a strict program, independent of the genome. According to this point of view, the genome is merely a passive partner in the first stages of development, and intervenes only subsequently to define the secondary characteristics of the organism. The general plan of the organism is supposedly fixed in the egg, not in the activity of genes during these first phases.

A number of precise examples will show that this apparent inactivity of the genes in the early phases of development is an illusion. First, it is observed only in some animals. In others, such as mammals, the genome intervenes very early in the construction of the organism. Second, even in animals where embryonic genes begin to function only later, the first stages of development are still entirely controlled by maternal genes, which have organized the structure of the egg and have prepared it for the first steps of development.

At the cellular level, the idea that genes are only tangentially involved in the replication of life flows quite naturally from the complex process of cell division. A cell is a highly organized structure, composed of a nucleus that houses the chromosomes and of a cytoplasm that contains a number of organelles. Each of these organelles has a role: the mitochondria make energy for the cell, in plants the chloroplasts capture light energy, the endoplasmic reticulum and the Golgi apparatus modify and guide proteins that are either in the membrane or are secreted. When cells divide, none of these organelles completely disappears: they fragment so that each daughter cell has a complement of each type of organelle. Observations of most kinds of cells thus suggest that the organelles have their own replicative capacities.

There is an element of truth in this. A basic element of the cellular structure—the membrane—cannot be created out of fundamental components: a membrane always derives from another membrane. This does not imply that a membrane cannot be modified by the action of genes—that it cannot receive new lipids or new proteins which give it its structural and functional characteristics, making it a membrane in the endoplasmic reticulum or the Golgi apparatus. But ever since the origin of life, a cell has always been born of another cell, and a membrane is always born of another membrane: the isolation of an internal medium from the external environment, via the formation of a membrane, probably constitutes life.

Furthermore, some of the organelles present in the cytoplasm—mitochondria and chloroplasts—were, in the distant past, independent organisms that were subsequently captured by other cells and now live in symbiosis within them. Despite no longer having an independent existence, they have retained a part of their genome that codes for some of their structures. There is indeed a form of heredity that is not linked to the nuclear chromosomes, even if it is of only limited importance and based also on a sequential heredity.

In the case of the other organelles (endoplasmic reticulum, Golgi), the fact that they are conserved during cell division does not necessarily mean that their components can be assembled only on a pre-existing organelle of the same nature. However, given that the *ab initio* process is slow, there was probably a strong selective advantage in favor of inherited mechanisms for the production of these organelles. The heredity of organelles is thus not another form of heredity in competition with genetic heredity. It is merely the result of the cell's parsimony, by which new organelles are derived from pre-existing organelles, rather than recreated at each cellular division.[17]

The difficulty in knowing whether genes contain all the information necessary to produce a given form is also encountered at the most elementary level of organic functioning, when the protein chain folds in order to acquire its function. Biologists have been studying this phenomenon for nearly 40 years, using the most sophisticated techniques that make it possible to follow events that take place in extremely brief periods of time. And yet in most cases they are unable to predict what form a given protein will adopt.

The fact that a precise, perfectly functional form emerges spontaneously from what appears to be a random sequence of amino acids is somewhat disconcerting. This explains why scientists were so pleased at the recent discovery of molecular chaperones, which fix themselves onto proteins while they are being synthesized and favor folding. Some thought that these chaperones would turn out to be the matrix that gives proteins their final form. Unfortunately, these hopes were dashed. Molecular chaperones are now known to play a passive role: they merely prevent the emerging proteins, as yet unfolded, from interacting with one another to create formless and inactive aggregates.[18] They are simply there to assist the spontaneous folding of proteins. There are no microscopic demons that give proteins their form; or rather, these demons do exist, but they are the physical laws of chemical kinetics and thermodynamics.

According to our current knowledge, there is only one example of well-documented nongenetic heredity, in which an organic form can create a similar form without recourse to the linear information contained in a nucleic acid molecule: that of prions.[19] The very name "prion" shows that despite a great deal of detailed research these pathogenic agents have been found to be composed only of proteins. Prions are responsible for degenerative diseases of the nervous system in humans and animals, such as Creutzfeldt-Jakob disease (CJD) in humans and mad cow disease (bovine spongiform encephalopathy, or BSE). These diseases can occur apparently spontaneously, as in most cases of CJD, or following oral infection, as in BSE—which was propagated by giving cows feed prepared from the corpses of contaminated animals—and in the human new variant of CJD—which results from eating contaminated beef. CJD has also been transmitted to children via the injection of growth hormone prepared from the pituitary glands of people suffering from CJD.

These diseases are apparently caused by the transformation of a normal cellular protein, called the prion protein, into a pathogenic form that aggregates and causes neurons to die. This transformation does not imply that the prion protein is stably modified but that there is a small change in the way that the polypeptide chain folds.[20] The formation of the pathogenic form is favored by mutations in the gene coding for the prion protein, which is present in those families where these diseases are hereditary. What distinguishes so-called prion diseases from other neurodegenerative processes where proteins are accumulated in an abnormal form, such as Alzheimer's disease, is that the pathogenic form of the prion can favor the conversion of the normal prion protein into the pathogenic form, thus activating the reproduction of its own form. Furthermore, this pathogenic form is highly resistant to all denaturing treatments, thus making contamination possible via feeding, for example.

This model of such diseases, the protein-alone model, is widely accepted. None of the competing models, which suppose, for example, that the prion is a virus and that it contains its own genetic material, has yet been supported by decisive experimental evidence. However, despite the fact that the

Nobel prize was awarded to Stanley Prusiner for the development of the protein-alone model, this model is weak. It cannot easily explain the existence of several variants of the pathogenic agent which have been demonstrated in a number of studies: these observations imply that there are different stable forms of the pathogenic form, each of which can favor the conversion of the normal prion protein into its own form. Specialists in protein structure—who generally admire the ability of proteins to adopt a precise structure—find this difficult to swallow.

If we accept for a moment that the protein-alone hypothesis will be confirmed in the coming months or years, that would mean that one protein can modify the form of another by transmitting its form. There would thus exist, at the molecular level, a heredity of form that was distinct from genetic heredity. But would this shake the foundations of molecular biology—the role of genes in the functioning and reproduction of organisms?

However serious the prion might turn out to be for public health, from a biological point of view it is merely an anecdote. The fact that one type of protein, the properties of which are not yet fully known but which are probably very unusual, can reproduce its form (or forms) without recourse to genes—or to be more precise, which can transmit in a contagious manner a deformation of its own form—should not lead us to conclude that all proteins, or even an important number of them, can do the same.

There are good experimental reasons for thinking that prion-like phenomena are very rare, indeed marginal. The genetic processes of models such as the fruit fly and yeast have been studied intensively, including those that did not fit in with the received wisdom, and based on these results we can be pretty sure that protein heredity is a rare phenomenon. Some cases have been described in yeast and one in a fungus.[21] If confirmed, self-replication of protein structure will be limited to a very few cases—more than enough to scare us all out of our wits, no doubt, but not enough to invalidate the question posed at the beginning of our investigation: how do genes control the development and functioning of organisms?

CHAPTER THREE

INVESTIGATING WHAT GENES REALLY DO

The previous chapter explained that genes are directly or indirectly responsible for the synthesis of all the fundamental components of an organism. We have seen that nongenetic heredity plays only a small role. Nevertheless, we have not got very far in our quest to answer the question, "What do genes really do?" To go further, we will have to adopt an experimental approach, dealing with matters case by case, gene by gene.

The fact that genes play a role in the life of the organism at two different moments—once in normal cellular life, the other when they are transmitted to the next generation at reproduction—means that two diametrically opposite approaches can be followed. In the first, the starting point is a gene for which we know, or think we know, the function in isolated cells *in vitro,* and we aim to determine its function *in vivo* by substituting a nonfunctioning version of this particular gene in living animals. In the second case, the starting point is a character that is known to be under genetic control, and we look for the precise function of the gene(s) involved. These are two opposite paths, but they should meet.

KNOCKOUT EXPERIMENTS

In this approach, after biochemical study of a protein in a test tube has shown that it is involved in a given process, the gene that codes for the protein is then modified in order to alter the protein's function (in most cases, to simply cancel that function). The consequences of this modification on both the development of the organism and its functioning are then studied. Using this technique of targeted gene inactivation ("knockout") in the mouse, researchers have, in the space of a few years, accumulated a massive amount of data, often unexpected, always interesting.[1]

In knockout experiments, the gene being studied is not directly modified but is rather replaced by a modified or inactivated copy. This cannot be done directly on an embryo. If an altered form of a gene is injected into an embryo, it will generally be integrated randomly into the genome and will replace the native copy of the gene only in a small fraction of embryos. Gene substitution therefore involves two stages. First, the modified copy of the gene is injected into embryonic stem cells, and by a series of experimental tricks those cells where the experimental copy of the gene has replaced the native copy are selected. These embryonic stem cells are then injected into a mouse embryo. They mix with the embryo's cells, participate in the formation of the adult tissues and, in particular, in that of the germ line. Some of the offspring of this first "mosaic" mouse will come from germ-line cells in which genetic substitution has taken place: thus, in every cell of their body these second-generation mice will have one normal copy of the gene and one altered copy. By simple mating crosses, it is possible to produce animals with two copies of the modified gene and thus to see the effects of the modification.

MAPPING GENES

The second approach flows from the function of genes in reproduction. The organism studied could be an animal, but it is often a human being. The character being studied is often a disease or an anomaly, but it could also be a physical or even a psychological character. However, the fact that a character can be transmitted does not necessarily imply that it has a genetic basis: it can equally well be transmitted by environmental factors. For example, children with parents who have tuberculosis are more likely to catch the disease than other children even if, genetically, they are no more susceptible to the disease than children in a disease-free family.

If transmission across generations is indeed genetic, then the gene(s) responsible for the character(s) can be mapped. The principle involved is to see whether, over the generations, the character remains associated with a genetic "marker" that has already been localized in the genome. On condition that this co-transmission can be studied in a sufficiently large number of families, and that the trait is not controlled by too many genes, the gene(s) involved can generally be localized at a particular point on the chromosome, through the use of genetic maps.

Two different situations can exist in which genes are associated with a disease. (Because diseases are generally clearly defined, it is easier to study the example of genes associated with diseases, but the principle also applies to genes associated with other characteristics of the organism.) In the first case, a mutation in the gene induces the development of the disease in most of the individuals carrying the mutation. In the second case, the presence of a particular form of the gene increases the probability that the individual will develop the disease when other required genetic or environmental factors are also present. In the former situation, the forms of the gene that are linked to the disease are relatively rare in the population, whereas in the latter case they may be relatively frequent.

In the first case, we generally speak (mistakenly, as we will see) of "disease genes," whereas in the second case the genes are described as genes that produce a susceptibility or a predisposition to develop the disease. In theory there is no clear divide between the two kinds of gene, but in practice there is an obvious distinction between rare, familial genetic diseases and widespread diseases that show no clear hereditary transmission but in which it is nevertheless possible to show the existence of susceptibility genes. In this latter case, there are generally not one but several forms of genes that contribute quantitatively to the probability that the disease will develop. Because there are so many such genes, and because each one only partially contributes to the development of the disease, it is clearly more difficult to localize them. Similarly, the genes involved in physical or psychological characteristics are generally multiple, and localizing them presents the same difficulties as identifying the genes involved in disease susceptibility.

The next step is the most difficult, because a genetic localization, even though relatively precise, is nonetheless extremely vague at the molecular level. The zone pin-pointed often contains dozens or even hundreds of genes, all potentially responsible for the character—"responsible" simply meaning that variation in the structure of the gene induces a variation in the observed character.

The final isolation of the gene is the result of one of two possible strategies.[2] The first is simply to see if, among the genes present in the zone described, some might not be good candidates simply on the basis of their function or the fact that they are expressed in the appropriate tissues. If this is the case, researchers concentrate their attention on such gene(s) and try to see if there are different alleles, some of which might be specifically associated with the character under study. The other strategy, used in particular in the case of familial genetic diseases, is to see whether some individuals that express the character have a chromosomal deletion—a loss of a small number of genes—or a chromosomal rearrangement in the region of the genome being studied. If this is the case, the problem is to find which gene(s) are deleted or modified by the rearrangement.

Once the gene is localized and sequenced, it is possible to determine a "virtual protein," the function of which can be suggested on the basis of similarities with previously described proteins. In general, it is possible to tell from the sequence whether the protein will be inserted into a cell membrane, act as a transcription factor, or have an enzymatic activity. These indications are important in order to predict the role of the normal protein.

In the case of studies on humans, the definitive proof that the isolated gene is really involved in the character under study may be provided by replacing the corresponding mouse gene (humans and mice are sufficiently close for there to be a corresponding gene in virtually all cases) with a mutated form and looking for any effects of such a substitution. This will work only if it is possible to compare the properties shown by the mutant mouse and the characters shown by humans with the same genetic alteration. In the best case—where the animal expresses the character shown by humans—we still have to understand how the mutation produces its effect.

SATURATION MUTAGENESIS

A third approach, possible only with animals, is a variant of the second. One starts with a given character and then, after having subjected the animal to a treatment that increases the natural mutation level, tries to isolate a large number of mutant animals in which the character is abnormal. The next step is identical to that described above, except that the genetic localization takes place in a model organism and not in humans. The only precondition for carrying out such a study is that the modification of some genes will alter the character being studied. The opposite would in fact be rather surprising, given that genes are responsible for the synthesis of all the molecules that make up the organism. Less restrictive than the second method, which is based on the isolation of natural genetic variants, this third approach, called saturation mutagenesis, aims to identify all the genes involved in a given character.

Each of these three approaches has its limits and its difficulties. Each is appropriate for different levels of partially distinct phenomena, and each answers the question "What do genes really do?" in a different way. Chapters 4–6 outline some results of these different approaches and show how they shed light on the question. In Chapters 7 and 8 these analyses will be used to tackle directly the role of genes in the construction of the organism, in aging, and in the control of behavior.

CHAPTER FOUR

GENES THAT CAUSE DISEASES

The first example of a disease with a genetic basis that I will discuss is sickle-cell anemia, characterized 50 years ago and caused by a mutation in a single gene. This disease has been intensively studied and can tell us a great deal about what genes really do. The name of this disease derives from the sickle shape of victims' red blood cells in the absence of oxygen. These deformed red blood cells are more fragile than the normal kind and they block tiny capillaries, leading to underoxygenation of body tissues (often accompanied by intense pain) and anemia.

The hereditary transmission of sickle-cell anemia was first described in 1949 by James Neel.[1] In the same year, the American chemist Linus Pauling showed that the disease was associated with a modification in hemoglobin, the molecule that transports oxygen in the blood.[2] The hemoglobin protein extracted from patients with sickle-cell anemia had an electric charge that was different from that of the normal version. Eight years later, Vernon Ingram and Francis Crick showed that this difference was due to the replacement of a single amino acid in the protein. This change leads to the aggregation of hemoglobin at low oxygen concentrations, and thus to the deformation of the red blood cell.[3]

The first lesson we can draw from the study of this disease is that the gene which, when mutated, causes sickle-cell anemia is not "the sickle-cell anemia gene" or "the gene for sickle-cell anemia," as is often stated in popular science books and articles.[4] The relevant gene is simply the one that normally codes for the β hemoglobin chain. The effect of the mutant form has nothing to do with the normal function of the protein, which is to fix oxygen. To understand this point, imagine another mutation in the same gene that affects the fixation of oxygen onto the hemoglobin molecule (several such mutations in fact exist). Such a mutation would lead to poor tissue oxygenation and would be clearly linked to the normal function of the protein. But this is not the case for sickle-cell anemia. The main symptoms of the disease are caused by abnormal polymerization of mutant hemoglobin at low oxygen pressure. The only logical link between the observed anomaly and the normal function of the protein is that the defect is the result of the abundance of hemoglobin in red blood cells. In addition, red blood cells require a very specific shape to move through narrow capillaries, and if their shape is distorted, movement will be impeded. If hemoglobin was not so abundant, or if the red blood cell was an immobile cell, the mutation would not lead to any pathology whatsoever.

Nevertheless, the fact that there is no logical link between the normal function of the protein and the effect of the mutation does not mean that it is wrong to explain the totality of symptoms, as I have just done, taking the mutation as the starting-point. It is quite legitimate to unravel the causal chain from the mutation through polymerization of hemoglobin and deformation of the red blood cell to arrive at anemia.

And yet, despite this causal chain, the disease is not under strict genetic control. Depending on individuals and populations, the same mutation can induce benign or lethal effects.[5] For example, in Central Africa, children with sickle-cell anemia die very young; mortality is so high that for some time it was thought that the disease was absent from these populations, since no cases were found in adults. On the other hand, in Southern India, many people with sickle-cell anemia survive quite happily into adulthood. This extraordinary variability in the gravity of the disease is due to the fact that the symptoms depend on the degree to which the hemoglobin β chain is expressed, and on the expression of other (minor) forms of hemoglobin, as well as on the possible re-expression of a fetal form of this protein, all of which can partially compensate for the effects of the abnormal form. The gravity of the disease will also depend on other red blood cell proteins that can modify the intracellular environment, as well as on the proteins present in the blood or on the surface of the cells that line the capillaries, all of which interact with the deformed blood cells. Although sickle-cell anemia is a genetic disease due to a simple mutation, in its expression and its evolution it is a multifactorial and multigenic disease.

The same conclusions could have been drawn from the study of phenylketonuria, a genetic disease that is caused by a deficiency of the enzyme that degrades the phenylalanine found in many foods. The absence of this enzyme leads to severe mental retardation, but these effects can be prevented following a diagnostic test made at birth and the provision of phenylalanine-poor food. As first noted by Lionel Penrose in 1946 and repeatedly confirmed since, it is hard to predict the gravity of the phenotype, even given a precise knowledge of the genetic lesion.[6] Perhaps it would be wiser to describe phenylketonuria as an environmental disease rather than a genetic disease, precisely because the phenotype is not seen if food sources are low in phenylalanine.

A similar situation is found in cystic fibrosis.[7] This disease, which is due to a mutation in a single gene, is the most frequently observed genetic disease in European populations. It results from the production of a thick mucus that blocks both the pancreatic secretory channels and the respiratory passages. The gene, which was cloned without any *a priori* ideas as to its function, codes for a chloride channel—a protein which, when it is inserted into a membrane, regulates the passage of chloride ions. This discovery made it possible to understand the origin of the disease. By blocking the passage of chloride ions across the membrane, mutations in the gene modify the composition of the fluids secreted by the cells and thus alter the composition of the mucus. In this disease, the primary deficit in the passage of chloride across the cell membrane is directly linked to the normal function of the protein. On the other hand, the main symptoms, which result from the blockage produced by the thick mucus in respiratory and secretory channels, are relatively indirect consequences of the initial deficit.

On closer investigation, the relation between the gene mutation and the phenotype (the symptoms of the disease) is even more complicated. The most widespread mutation in the European population—the loss of a phenylalanine amino acid at position 508 in the polypeptide chain—is only indirectly responsible for the chloride channel deficit. In a way we still do not fully understand, this mutation leads to the specific degradation of the protein. As in the case of sickle-cell anemia, the gravity of the disease varies, even between individuals with the same mutation. Although pancreatic insufficiency is always observed, the respiratory effects vary in gravity and are probably largely controlled by other genetic or environmental factors. It should be added that, in men, this mutation can sometimes lead to sterility; indeed, in some cases this is the only symptom.[8]

The fact that the same mutation can produce different

syndromes is even clearer in other genetic diseases. Some mutations affect the form of the face by inducing premature bone fusion during head formation. These mutations often concern genes coding for a particular type of protein—growth factor receptors.[9] Growth factors are small proteins that stimulate cell division; a cell with a growth factor receptor can respond to these signals. The modification of one of these receptors can lead to four different syndromes. It might be thought that this receptor plays several roles and that, in each syndrome, a different part of the receptor, and thus a particular function, is affected. In fact, however, the same mutation can lead to different syndromes in different families.

The same syndrome can also be produced by mutations affecting different genes. These observations show that we should abandon the naive idea of a simple one-mutation, one-disease relation. Different mutations, different pathogenic pathways, can lead to the same symptoms. And conversely, the same mutation can lead to two different pathological states. How is this possible? Do other genes affect the final result? Certainly. Does the environment play a role during development? Perhaps. One day, it will no doubt be possible to predict that a given mutation, in the presence of a given form of another gene, or such and such conditions in the external environment, will induce a given syndrome, whereas the same mutation, in the presence of a given form of another gene, will lead to another syndrome. Our current difficulties will be partly dispelled and genetic determinism will apparently be put in its proper place. Nevertheless, this example shows that the landscape of potential forms—or at least of pathological forms—is limited, and that genes simply orient the organism to one of a number of possible pathways.[10] Genes act more like railroad switches than like the operator of a four-wheel drive vehicle who can go where he or she wants.

These examples show that the links between the symptoms of a disease and the mutations that are at the origin of the disease may be highly complex. This is not to say that it is impossible to explain the relation between an illness and a mutation, but that this relation is indirect, unpredictable, and of variable intensity. A series of further examples will illustrate and develop these remarks, by focusing on genetic diseases that affect the functioning of the central nervous system and of intellectual and cognitive abilities.

GENES INVOLVED IN NEUROPATHOLOGIES

A good example is Williams syndrome.[11] The way its genetic bases have gradually been discovered reveals the difficulties of interpretation encountered by geneticists, as well as the pitfalls that await them. The first patients to be studied all had a relatively large deletion on one copy of chromosome 7 (over 500,000 base pairs, more than enough to contain several genes). The clinical picture of the disease is complex. Patients show a shrinking of the major arteries, which slows blood flow; facial anomalies; a pelvic hernia; and an elevated calcium level in the blood . They also suffer from mental retardation, with very precise behavioral and cognitive characteristics: their auditory memory—in particular their linguistic memory—is altered, they find it difficult to recognize and reconstruct complex objects from separate elements, and they are highly gregarious.

A study of a family affected not by Williams syndrome but simply by a shrinkage of the aorta showed that this particular anomaly was linked to a modification of the elastin gene, which is situated in the same region of chromosome 7. Elastin is an abundant protein which, as its name indicates, helps make tissues elastic. Subsequent investigations showed that in 90 percent of individuals with Williams syndrome the elastin gene was deleted, but this deletion was responsible for only the facial and arterial anomalies and for the pelvic hernias. It was not responsible for the cognitive defect. It thus appeared probable that the cognitive deficit was the result of the modification of another gene situated in the same region of chromosome 7. The chromosomal deletion is important and (probably) includes a large number of genes. At the time, sequencing such a large fragment was difficult. Furthermore, it would not have *directly* revealed the gene involved in the defect.

The geneticists who were studying this disease therefore tried to work out what the gene might do.[12] As we will see, studies of memory had shown the importance of a particular class of enzymes, the protein kinases, and even of a particular sub-class of these kinases, the tyrosine protein kinases. The sole function of protein kinases is to modify the structure and activity of other proteins and enzymes by adding phosphate groups to specific amino acids, such as tyrosine. There are many kinds of these kinases, and they play a very important regulatory role within cells. Using the DNA amplification technique, scientists investigated whether the genomic fragment that was missing in patients with Williams syndrome contained a sequence corresponding to a tyrosine protein kinase. And indeed, such a sequence was found very close to the elastin gene. Furthermore, this gene also carried a sequence that had already been found in several genes involved

in the development of both the nematode worm and amphibians.

Analysis of a patient who suffered only from the cognitive
defect associated with Williams syndrome showed that the
gene coding for this tyrosine protein kinase had a specific
modification. During development, this gene is expressed in
the brain. It thus seemed very probable that the malfunction-
ing of this gene, and of this gene alone, was responsible for the
cognitive defect associated with Williams syndrome. However, recent results suggest that this conclusion could be too
hasty, and the situation remains unclear.[13]

This example shows how careful we must be in relating a
given genetic disease to the observed modification of a gene.
If the provisional conclusions were to be fully accepted, it
would also show that the dysfunction of a single gene, coding
for a protein with a simple function, can block or at least alter
a complex cognitive function. It would, of course, be utterly
wrong to conclude that this cognitive function is due to this
gene, that it can be explained by the existence of this gene. All
this series of studies shows is that, when the product of this
gene is altered, the function in question cannot take place
normally. We will return to this question at the end of the
chapter, after having described two other genetic diseases that
are serious both in terms of the number of patients affected
and the behavioral modifications involved.

Huntington disease is a "dominant" genetic disease—it is
produced by the modification of only one of the two copies of
the gene involved that are present in each individual. Its
penetrance—the probability that if the mutation is present,
the disease will develop—is high. The disease generally appears after the patient has reached 40 years old, and it leads to

motor disorders, seizures, depression, dementia, and, within a few years, death. Current techniques make it possible to predict with a high degree of certainty whether a healthy adult individual will succumb to this particularly incapacitating disease, for which there is no cure.

Cloning the gene involved was a long and arduous task. The function of the large protein coded by the gene—huntingtin—remains a mystery. However, we do know that it plays a fundamental role: in mice, inactivation of the gene leads to death during development.[14] The genetic modification associated with the disease is not a simple mutation but the expansion of a monotonous repeated sequence of a single amino acid, glutamine, within the normal protein. The longer the polyglutamine sequence, the earlier the appearance of the disease.

Interestingly enough, long sequences of glutamines have also been observed in other neurodegenerative diseases with a genetic origin. However, for each disease, the protein within which the number of glutamines is increased is different, and specific symptoms result. In a particularly revealing experiment, transgenic mice were produced that carried the part of the gene containing these polyglutamine sequences, which had been isolated from a patient with Huntington disease.[15] The transgenic animals expressed the abnormal movements and seizures characteristic of the disease in humans, even though no disease resembling Huntington's has ever been described in mice. Like human patients, the mice also showed a reduction in brain mass. At the microscopic level, the nuclei of the nerve cells showed analogous anomalies to those seen in humans. The experiment was reproduced with a genetic construct allowing researchers to control the activity of the gene which was introduced. As previously, the gene's expression appeared to induce symptoms of the disease. When its expression was then turned off, the symptoms disappeared and the mice returned to their normal state. Such a surprising observation strengthened the link between the production of polyglutamine chains and the appearance of the disease.[16]

These experiments suggests that every symptom of the disease is simply linked to the synthesis of these long chains of polyglutamines, with no relation to the normal function of the protein coded by the gene. These polyglutamine chains may induce the death of the nerve cells in which they are synthesized. We know that these long chains of polyglutamine aggregate inside the cells, although aggregation does not seem to be directly correlated with cell death.[17] Despite the 'mechanistic' origin of the disease, its age of appearance[18] as well as its specific form can be influenced by the environment and can differ between two identical twins[19]: genetic determinism and environmental effects are not mutually exclusive.

The final disease that we will examine at the molecular level is Alzheimer's, which is responsible for many cases of senile dementia. In some families, this disease clearly has a genetic origin, with an early onset of symptoms. It is linked to extracellular deposits of aggregated proteins in the brains of affected individuals. The earliest kinds of deposits, which parallel the development of the disease, contain a peptide called β-amyloid, a truncated form of a protein that is normally present in nerve cell membranes. When this peptide is added to nerve cells, it provokes an abnormal oxidation of lipids, changes in glucose metabolism and, ultimately, cell death.[20]

Studies of hereditary early-onset forms of Alzheimer's disease have found mutations in the β-amyloid protein precursor gene in only a very few cases. On the other hand, two other genes are responsible for around 50 percent of those cases of early-onset Alzheimer's disease. The protein products of these genes are called presenilin 1 and presenilin 2. The sequence of these proteins revealed nothing about their potential function. Their localization within the cell, on the other hand, was more informative. They are present in the endoplasmic reticulum and in the Golgi apparatus, which suggests that they might be involved in the transport and maturation of membrane and secreted proteins.

More precise ideas as to the function of presenilin came from studies of the nematode *C. elegans*. This small worm was chosen by Sydney Brenner, one of the key actors of the molecular revolution of the 1950s, as a model organism through which to understand the genetic mechanisms that guide development and the formation of the central nervous system.[21] Modest in its feeding requirements, able to reproduce rapidly and without a sexual partner (both of which make genetics research easier), the worm is sufficiently complex for experimental results from studies of its development and of the formation and functioning of its nervous system to be applicable to more complex organisms, including humans.

Today, we know more about *C. elegans* than about any other multicellular organism. All the cell divisions that lead to the formation of the adult organism have been recorded. The connections between its different neurons have been described down to the last detail. Many mutations affecting the development and behavior of the worm have been isolated. The nematode has become a model system for the study of aging and cell death.

Even if the development of *C. elegans*—and, in particular, the development of its nervous system—is far more rigid (and simpler) than that of the human brain, the example of presenilin shows how results obtained with this organism can be useful to understand the situation in humans. One of the families of developmental genes that have been most widely studied in both nematodes and mammals is the *notch* genes. The original *notch* genes were discovered early in the history of *Drosophila* genetics: in their mutated form these genes produces small notches in the fly's wing veins. This relatively minor phenotype turns out to be merely the final step in a fascinating pathway that has profound implications for development in most animals, including humans. The *notch* genes code for proteins located in the cell membrane that are involved in interactions with neighboring cells. They thus control the cell's "decision" to opt for a particular developmental fate. By looking for gene products that interacted with the gene products of *notch-1*, nematode specialists isolated a gene that closely resembled that of presenilin.[22] This association between presenilin and Notch-1 was subsequently confirmed in mice.

On the basis of these results, one might imagine that the function of presenilin is to interact specifically with the product of *notch-1*, in order to control certain early stages of development. However, given that *notch-1* codes for a membrane protein that is obliged to pass via the endoplasmic reticulum and the Golgi apparatus before reaching the membrane, a simpler hypothesis, so straightforward as to be almost banal, is that presenilin is a protein that is essential for the transport or the maturation of proteins inside this organelle. According to this hypothesis, this protein's high degree of conservation through evolution would be the consequence of the conservation of a nonspecific but essential function.

This hypothesis has been confirmed: if mutant forms of

presenilin are introduced into mice, the cleavage of the β-amyloid protein is altered,[23] and presenilins were recently shown to bear this cleavage activity.[24] Alzheimer's disease, which is responsible for devastating effects on memory and behavior, could turn out to be the result of a problem of protein maturation. However, mice carrying mutations only in presenilin genes are healthy and show no signs of developing the disease: only those animals that also carry a mutation in the transgene coding for the amyloid precursor protein show an Alzheimer-like phenotype.[25]

These mouse experiments show that the relation between the formation of the peptide and Alzheimer's disease is in fact highly complex. It may even be the case that the erroneous cleavage of the β-amyloid protein is merely a secondary consequence of a more serious change in the endoplasmic reticulum, which leads to cell death; it has recently been shown that cell death is accompanied by the production of a family of enzymes that can lead to the formation of the β-amyloid peptide.[26] The formation of the β-amyloid peptide could thus turn out to be more a consequence of the disease than its cause.

Finally, as well as those genes which, when mutated, lead to the development of early-onset forms of familial Alzheimer's (that is, genes coding for presenilins), there are also "susceptibility" genes that appear to be associated with late-onset forms of the disease, which are also the most common. These genes, *ApoE* and *A2M,* are present in several forms, some of which are undoubtedly more frequently found in people with the illness.[27] The mechanism by which these genes, which are apparently involved in lipid metabolism, play their role in the onset of Alzheimer's disease remains unknown. Whatever the case, it is interesting to note that these genes are of a different kind to those involved in the early-onset familial forms of the disease.

SOME PRELIMINARY CONCLUSIONS

We have not talked about the genes that might be involved in psychiatric diseases, such as schizophrenia or bipolar disorder. The isolation of such genes has been so far unsuccessful. The main difficulty is in the definition of the diseases: the present categories enclose different pathologies, which makes the work of geneticists impossible. However, this brief overview of genetic diseases emphasizes a number of interlinked points:

- The relation between a given disease and the modifications of the gene involved is highly variable.
- The disease may be explicable on the basis of the normal gene function, or it may have no apparent link with this function.
- The effect of mutations may be clearly determined independently of other genes and of the environment, or it may be so totally dependent on these factors that they have more of an impact than the mutated gene itself.
- A given disease may be linked to different kinds of genes depending on whether it is inherited or not, and whether it has an early or a late onset.

It could be argued that this merely suggests that the approach outlined in this chapter is profoundly mistaken, and that it is pointless to try to understand gene function on the basis of such pathological malfunctions. However, this method in fact forms part of a long medical and biological tradition that studies pathology and disease in order to understand normal functioning. William Harvey, who discovered the circulation of the blood, argued that it was possible to discover the laws of nature by the careful study of rare diseases.

In reality, the plurality and complexity of the observations described here are of the utmost importance. First, they show the diversity of genes and the mechanisms of gene action. They make us think about the link between gene structure, gene products, and phenotype (the observed character, be it in normal or mutated individuals).[28] Very often, the gap between the observed anomaly and the fault in gene structure or function is so great, and the number of intermediary factors so large, that the defect cannot be predicted from the gene mutation, or even from its initial cellular consequences. In the case of Huntington disease and probably also Alzheimer's, neuronal death, which is partially responsible for the symptoms, is not the "natural" consequence of gene malfunction, nor even of the formation of the protein deposits observed in these diseases. As we will see, cell death is the consequence of the activation of a complex program which requires that the doomed cell decodes signals from both inside itself and from the environment in a way that is still poorly understood.

We should thus be wary of hasty explanations and simplistic schemas. It could be argued that because the partial absence of tyrosine-protein kinase seems to be linked to the cognitive defects seen in Williams syndrome, this enzyme might play a fundamental role in such cognitive functions. Such a simple hypothesis may not in fact be appropriate. The data—which are still far from complete—suggest only that the low level of this enzyme leads to a certain number of cognitive defects. However, the action of the enzyme could be direct or indirect. It could, for example, be necessary for the construction of cerebral structures that are required for carrying out the cognitive function. In favor of this hypothesis, it should be noted that the cognitive defects found in Williams syndrome are analogous to those found when there is a lesion of the posterior right parietal region of the brain.

The problem of the link between a deficit and a normal function is not only posed in the case of genetic diseases. Since the nineteenth century, scientists have tried to localize certain superior cognitive functions in the brain by linking alterations of these functions to changes in the corresponding cerebral regions. This approach has given rise to many debates, at least as many as (if not more than) those provoked recently by research into gene function. The opponents of the "localizationists" have always advanced the same argument: we cannot conclude that a cerebral function is localized in a particular brain region simply because this function changes each time the brain region is physically altered. All this tells us is that the region is necessary for the realization of the particular function. However, now that magnetic resonance imaging (MRI) and positron emission tomography (PET) make it possible to observe directly the activity of different brain regions while a particular cognitive task is being carried out, it has become clear that the brain regions activated are precisely those that, when altered, prevent the task from being executed.

This history lesson from the recent past is also useful for the study of genes. A gene that is mutated when a particular function is impaired may indeed turn out to play an essential role in the normal expression of the function—even if this in no way implies that the gene involved is *the* gene for this particular function.

CHAPTER FIVE

KNOCKOUT SURPRISES

Knocking out a gene involves replacing the normal copy of the gene with an abnormal copy, thus giving rise to an inactive protein product or to no product at all. The knockout technique makes it possible to carry out selective, directed mutagenesis.[1]

This technique is particularly useful when the function of a gene is entirely unknown. For example, a gene may have been isolated because certain of its mutations are associated with the formation of cancerous tumors. However, such observations may tell us little about the function of the gene in the normal life and development of the organism. Inactivating the gene makes it possible to see in which tissues and organs its action is necessary. Conversely, when the product of a gene has been sufficiently studied and its function *in vitro* or in isolated cells has been fully described, it may seem unnecessary to verify the function *in vivo* by a knockout experiment. However, knockout experiments (which have been carried out on about 1,000 genes) have produced more surprises than even the most enthusiastic partisans of this new technique expected.[2]

Having examined many Ph.D. theses, I have come to the conclusion that many young biologists suffer from a particular psychological trait which I call "knockout repression." Ph.D. theses often involve characterizing the function of a given protein. In many cases, parallel to the student's work another laboratory has knocked out the gene coding for the protein. One might expect that the Ph.D. student would be pleased to learn the results of such an experiment, because they could be included in the thesis and compared with results obtained *in vitro*.

After studying a large sample of Ph.D. students, I have come to a very different conclusion. The knockout experiment is always cited (the candidate cannot run the risk of being criticized for not knowing of it), but the results are described rapidly, in at most two or three lines, at a point that has been carefully chosen to guarantee that the inattentive reader will not notice. There are two complementary explanations for this behavior. The first is that, in many cases, the results of the knockout do not confirm the hypothetical function of the gene established on the basis of *in vitro* experiments. The second is that the knockout experiment, which is very difficult to perform, is given greater weight than the *in vitro* experiments. This unconscious valorization is also partly a reflection of the importance given to a genetic approach as compared with, say, a biochemical approach.

This chapter will deal with the first of these points—the difference between the results of *in vivo* and *in vitro* studies, that is, the surprises produced by many if not all knockout experiments. In some cases the knockout may have no effect whatsoever, despite the fact that the gene codes for a protein that is thought to be essential. Sometimes, the knockout can even improve the performance of the mutated animal![3] In other cases the knockout can have effects that are completely different from those that were expected.[4] Then there are examples where the inactivation of the gene does indeed produce the predicted defect, but to a lesser extent than expected. Another possibility is that when a gene has several functions, only one of those functions is perturbed or rendered inoperative by the knockout. Finally, there are cases where the inactivation of a gene does indeed lead to the predicted effect, but these examples are relatively rare. It is unfortunate that groups working on knockouts are not asked to describe the expected results before carrying out the experiment. The comparison of observed and expected, which cannot be done *a posteriori,* would reveal many of the problems associated with this technique.

In this chapter, as in previous ones, I will provide a wide variety of examples, with the aim of showing the complexity of observations rather than artificially simplifying them. We will then focus on knockout studies related to the development and functioning of the central nervous system and the realization of cognitive functions.

Some Lessons for Biologists

First we will look at some cases where a gene knockout has not overturned our previous understanding but rather has confirmed, completed, or slightly corrected it.

Over the last forty years researchers have made considerable progress in understanding the immune system—in describing the B and T lymphocytes, in characterizing the receptors that recognize foreign molecules, and in elucidating the mechanisms employed to deal with invaders. Initial investigations centered on the inactivation of genes coding for these receptors. The results were largely unsurprising, confirming the importance of these receptors in the immune response. However, these studies also showed that, in the absence of major components of the system, other components that had previously been considered to be minor elements could partly replace them. This functional compensation, which can occur either during the development of the system (shortly after birth) or during adulthood, is one of the most recurrent results found in knockout studies. Current studies of knockout genes that code for these minor components or for the many signaling factors that are exchanged by the cells of the immune system may hold some surprises. Up until now, however, these experiments have amply confirmed the patient work of the immunologists.

Rather than simply stopping a given protein from having the function attributed to it by *in vitro* studies, the knockout technique can result in its role being either extended or restricted. For example, there is a liver protein which is a receptor for dioxin,[5] a chemical used as a weed killer. (Dioxin was a major component of the infamous agent orange used by the U.S. army in Vietnam to defoliate the jungle, making it possible to observe enemy troop movements from the air.) Dioxin is processed by the liver, which contains a number of enzymes that are specialized in detoxification—the elimination of foreign components. However, in the case of dioxin and some other compounds, the action of these enzymes leads to the production of intermediate substances that are more reactive, and thus more dangerous, than the original substance. These toxic products may interact with DNA and produce mutations that will later provoke tumors to form.

These detoxification enzymes are not made by the liver all the time; rather, their synthesis is induced in response to the presence of foreign molecules in the external milieu. The dioxin receptor is a protein that can fix this substance, as well as a large number of other toxic substances such as benzopyrene (found in cigarette smoke). Once the toxic molecule has fixed to the receptor, the receptor activates the expression of genes coding for the detoxification enzymes. It might be expected that animals in which the gene for the dioxin receptor had been inactivated would be less sensitive to the presence of this substance. Although this is indeed the case, the mutated animals also present a series of unexpected anomalies: high early mortality, reduction of the number of white blood cells in the spleen and the lymphatic ganglions, and 50 percent reduction in the size of the liver, with major alterations to the bile ducts. This shows that the receptor plays an important role in the development of the liver and of the immune system. The function revealed by biochemical studies—the elimination of toxic substances—was only one of the receptor's roles, and probably not its primary function. With hindsight, this seems logical. It is simpler for an organism to reuse proteins and enzymes already present in order to produce a different function that is required only in highly specific circumstances, such as detoxification.

On the other hand, knockout experiments can show that a protein's function is more limited than one would expect on the basis of biochemical studies. For example, some enzymes are essential for the synthesis of prostaglandins, which have become the focus of attention since it was discovered that certain broad-spectrum medicines, such as aspirin, act by inhibiting the synthesis of these substances. Recently, many studies have shown that prostaglandins regulate the immune response, fever and pain. More generally, they act as quasi-universal modulators of the mechanisms that enable cells to adapt to changes in their environment or to the needs of the organism. They are also thought to be involved in cell division and cell death.

However, when one of the two genes that permit prostaglandin synthesis is destroyed, the affected animals show very few defects. Given that these two genes are not expressed at the same time in the same tissues, this somewhat bizarre result cannot be explained by invoking a gene-compensation effect. The only possible explanation is that the *in vivo* role of prostaglandins is in fact more restricted than had been deduced on the basis of *in vitro* studies.

The authors of an article in the journal *Cell* describing the results of these knockout experiments suggested that the question now raised was whether these mice suffer from headaches.[6] They were ironically underlining the ability of scientists to shift attention away from difficult problems. If only minor effects are observed instead of major ones, then the solution is simple: the minor question becomes a large one that has to be studied urgently.

A final example of a surprising knockout result will enable us to compare data on gene function from knockout experiments and those from the study of genetic diseases. Chapter 2 explained the role of the prion protein in neurodegenerative diseases such as Creutzfeldt-Jakob disease and mad cow disease. The protein-only model suggests that the mere change in conformation of this protein provokes the appearance of a transmissible pathogenic form. Inactivation of the gene coding for the prion protein supported this hypothesis, because mutated animals—unlike normal animals—no longer develop the neurodegenerative disease, even when major quantities of the pathogenic form of the prion protein were injected directly into their brains. But the surprising result is that the inactivation of the gene leads to relatively minor defects.[7] The animals are viable and present only minor neurological problems, often when the individual is quite old, on nothing like the scale of defects seen with the disease. This example shows, as we saw in the previous chapter, that the symptoms observed in a genetic disease can be linked to a particular form of the gene and thus of the protein present in patients rather than to the mere loss of normal function.

What are the lessons of these initial studies, taken from a range of biological disciplines? Sometimes such experiments reveal new functions. More frequently, however, they minimize the importance of the proteins being studied, and thus of the genes that code for these proteins. Three hypotheses can account for this apparent absence of function. If the protein or the gene has been studied in a different species from the mouse, it is possible that the two organisms, even if they are closely related, do not use the same genes to perform the same function. For example, there is a rare human disease called Lesch Nyhan disease which leads to an unusual kind of dementia involving self-mutilation in young children; it is rapidly fatal. This genetic anomaly corresponds to the inactivation of an enzyme involved in the biosynthesis of the constituents of RNA and DNA. Researchers have reproduced this genetic abnormality in mice. Mutated mice show no detectable pathology, suggesting that only humans use this metabolic pathway in the central nervous system to any important degree.[8] The lessons are that humans are not mice, and that the living world is above all characterized by its diversity.

The second hypothesis is simply that *in vitro* studies may have overstated the importance of the gene in question, perhaps as a result of that well-known human trait (particularly well-developed among scientists) of considering that what they do is extremely important. Other, more scientific reasons can also be put forward. For example, cells that have been isolated and grown in tissue culture might have different characteristics from those found in an organism. Given that *in vitro* studies are carried out on at most a handful of systems, it is possible that the choice of these systems may have biased the interpretation of the results, leading to an exaggerated role for a given protein (and gene) which, in the living organism, has a much less important function.

The final hypothesis is that the studied gene does indeed have an essential role, but that—*in vivo* much more so than *in vitro*—a deficiency in its function is compensated for by analogous systems. In particular, the organism has the whole of its development to overcome the deficit. The proven influence of other genes on the effect of a knockout supports this hypothesis.[9]

After this brief overview of the effects of knockouts, we need to focus on particular functions in order to better understand the role of genes. The rest of this chapter deals with genes that give the cell its form, enable it to interact with neighboring cells, and enable it to respond to environmental signals. These genes play an essential role in all cells, in particular nerve cells. Having studied these various intracellular signaling pathways, we will be ready to analyze in the next chapter the molecular mechanisms and the genes responsible for learning and memory.

CELL SHAPE AND INTERACTIONS

The growth of cell biology followed closely on the heels of the molecular revolution of the 1950s. Between 1970 and 1985 the highly reductionist ambitions of some molecular biologists that led them to try to explain all biological phenomena by the direct action of molecules were gradually abandoned, and the cell found itself restored to the place it had occupied since the nineteenth century, as the fundamental level of organization

of life.[10] Cell biologists have concentrated on studying those
mechanisms that enable the cell to direct its constituent proteins toward the organelles where they will carry out their
functions, and those that allow cells to interact with the external milieu and with other cells. Alterations in the latter mechanisms are involved in the formation of cancerous tumors.
This new cell biology uses the concepts of molecular biology
and the tools of genetic engineering but respects the organization and functioning of the cell. It interprets in molecular
terms functions that are strictly cellular.

The knockout technique has made it possible to test the
reality of mechanisms that had been suggested on the basis of
such studies. And again, the results of the knockouts are often
very surprising—perhaps even more surprising than those we
have just described.

A relatively simple case is the deficit in cell adhesion proteins—molecules in the cell membrane that stick cells together. Cell adhesion proteins play an important role in ensur-
ing close contact between epithelial cells, thus creating a
boundary between the inside and the outside of the organism.
Mammalian embryos form such an epithelium very early in
development. To do so they express an adhesion molecule
called E-cadherin. Knocking out the appropriate gene produces very clear results: without E-cadherin, the embryo fails
to implant into the uterus and dies rapidly.[11] Interestingly, the
deficit does not appear immediately, when the first cells start
to interact and the embryo compacts itself. Maternal E-
cadherin makes this first stage possible, but it is not present in
sufficient quantity to ensure the long-term formation of a stable epithelium. In this case, the same gene (and the same protein) are important both at the very beginning of development
and later during embryogenesis, but the protein is first pro-

vided by the maternal gene and then later by the embryo's genome. Thus, the strict division between maternal and genetic factors, beloved of certain biologists, is clearly artificial.

Another knockout case involves the extracellular matrix, which consists of proteins secreted by cells. In addition to providing external support for cells, this matrix orients and guides cells as they migrate during development. It might be expected that knocking out the genes that code for the components of the extracellular matrix would have a dramatic effect on the development of the organism. This is indeed the case for the genes coding for fibronectin or collagens. But it is not always so. For example, C-tenascin, a protein whose location at specific places in the amphibian embryo strongly suggests an active role in controlling the first wave of cell migrations, can be eliminated from the mouse without causing any problem.[12]

This finding could be interpreted by imagining that the different components of the extracellular matrix have partially redundant functions. Intergenic redundancy—two proteins contributing to the same function, so that the absence of one would be partly compensated for by the activity of the other—is one of the main discoveries that have been made with the knockout technique.[13] There are probably very few cases of perfect redundancy—two proteins with exactly the same function. But partial redundancy, where two proteins have a number of functions, some of which are unique, others of which are shared, is frequent.

Strictly speaking, redundancy needs to be distinguished from compensation, even if in experimental terms it is difficult to do this with the knockout technique. Compensation implies that the compensating gene codes for a different protein, with functions other than those of the altered protein, but that this gene will nevertheless compensate for the absence of the mutated gene.

Functional redundancy fits in with the partial or total duplication of genomes that has occurred over evolution, and the resulting existence of multigene families. The existence of redundancy had not been predicted by population geneticists—it was even considered highly unlikely. In theory, natural selection should have eliminated one of the two duplicate genes. However, studies of living populations have shown that this is not the case; and recently population geneticists have finally proposed a number of models that can explain the appearance and continued existence of redundancy.[14] The simplest hypothesis is that redundancy is never perfect and that natural selection continues to act on each gene. However, this raises some doubts about the predictive power of the models of evolutionary genetics!

In the case of C-tenascin, it is also possible that the results observed in transgenic mice do not in fact contradict the observations and hypotheses made in studies of amphibians but simply underline once again that, even in organisms which follow the same developmental pathway, nature has not always used the same elements, the same proteins, to realize the same function and the same objective.

Surprises were also in store when it came to determining the role of the components of the cytoskeleton which give cells their form and mobility. Each cell type contains specific intermediate filaments (so-called because of their size). The importance of these proteins in the form of cells and in the three-dimensional construction of organisms is suggested by the fact that they are absent from bacteria and from single-cell organisms with a nucleus. This indicates that they appeared relatively late in evolution. However, knockout experiments have only partly confirmed the importance of these proteins. The role of cytokeratins appears clearer: this class of intermediate filaments is specifically expressed in epithelial cells. It has recently been shown that there are several human mutations that modify these keratins and lead to skin diseases.[15]

There is another form of cytokeratin, cytokeratin 8, which is expressed in the early embryo. A mutation in the corresponding gene leads to early embryonic death. But this lethality does not affect all individuals, and some mutated mice can reach adulthood. The effect of this mutation—its penetrance—is thus not absolute and no doubt depends on other genes and other proteins present in the cells of these organisms. However, the network formed by cytokeratin has never been seen to be replaced by another type of intermediate filament in such animals.

Vimentin is also an intermediate filament protein, widely expressed in various embryonic precursors and in virtually all cells grown in culture. Knocking out the vimentin gene has no effect: the mice are viable and fertile.[16] None of the tissues in which this protein is expressed—sometimes very abundantly—shows any sign of alteration. None of the other proteins that participate in the formation of intermediate filaments is overexpressed in these animals to compensate for the defect: electron microscopy does not reveal the existence of another network of intermediate filaments in the mutant cells where a vimentin network is normally found.

Mutating the gene coding for the GFAP protein, which makes up the intermediate filaments found in astrocytes—cells present in the nervous system that form the immediate environment of neurons—had no major effect on development, survival, and reproduction of mutated animals.[17] Mutations merely led to limited alterations in the structure and function of the nervous system.[18]

What can we make of these results—demoralizing to some researchers who have spent years studying the function of these proteins, sometimes focusing on subtle modifications that were thought to explain functional regulation? As far as the intermediate filaments are concerned, no proof of any functional redundancy, or replacement of one type of absent filament by another, has ever been found. Two hypotheses, which have already been partly discussed, can be put forward to explain these results. Either the deficit created by the knockout is compensated for by the action of other cells that surround and mold the mutated cell, or the deficit exists but is not detectable under laboratory conditions. Some biologists think that the intermediate filament proteins might protect cells from deformations that they experience during stress. From this point of view, their absence would be problematic only in an adult animal that had already experienced such stress. This hypothesis seeks to justify the existence of the intermediate filament proteins: if their absence has no effect, their presence cannot have any selective advantage and they should not have been conserved over evolution. It is difficult to refute the hypothesis that intermediate filaments may provide a selective advantage under specific conditions. If no selective advantage is found, critics can always say that this was because the highly specific conditions under which the advantage is expressed were not reproduced in the experiment.

Many studies of the inactivation of genes coding for intermediate filament proteins have produced negative results. Such studies reveal the strategies used by scientists to cast their results in the most favorable light, come what may: to the nonexpert reader, the consequence is to hide the surprise that is often the immediate result of the knockout experiment. The absence of any obvious effect produced by the knockout of the gene coding for GFAP, the intermediate filament protein in astrocytes, is a case in point. The title of the article claimed that "Mice devoid of the glial fibrillary acidic protein develop normally and are susceptible to scrapie prions." The result, which was negative because it cast no light on the role of this type of intermediate filament, was given a positive spin by suggesting that astrocytes play no role in prion infections and that only neurons are involved in the development of such diseases. Unfortunately for the authors, recent studies show that in the absence of neurons, astrocytes can propagate the pathogenic form of the prion protein.[19]

SIGNALING NETWORKS

These studies of intermediate filaments are striking and disturbing, but it is not clear what lessons can be drawn. Results from knockout studies of intracellular signaling networks are both richer and more complex. These networks are used for simple but essential tasks such as metabolic control or cell division, and for highly complex tasks such as learning and memory. The study of such networks and their mutated versions is thus a key way of discovering how genes can be implicated in complex functions despite not being specific to these functions.

There are two important difficulties in presenting these data. The first is that the regulation networks and signaling pathways are highly complex, and the nonspecialist reader can easily get lost. The second is that the results of knockout studies in this domain are incomplete and still in progress—data are being accumulated in such a way that it is difficult to know how they fit together.

The simplest way to explain matters is to follow the pathway taken by signals within the cell. The signal comes from outside the cell, it interacts with the cell membrane, and it is then relayed to the inside of the cell, where it heads for the nucleus, either to activate or inhibit the expression of certain genes or to regulate cell division. Many extracellular signals, which are transmitted by hormones or growth factors and activate cell division, pass via a protein located on the inside of the cell membrane—the Ras protein. This protein plays such a central role in cell division that a simple mutation involving the replacement of one amino acid by another is a straightforward way of making any cells that carry such a mutation become cancerous. In about half of human cancers, one of the Ras proteins (there are three forms, each expressed in different tissues) is mutated, suggesting their importance both for the control of cell division and for the breakdown of this control.

The activity of Ras proteins is controlled by GAP proteins that interact with a large number of protein partners and are also the targets of many protein kinases, as would be expected if they play a fundamental role in the control of Ras protein activity. There are four GAP proteins in mice, one of which is present in a large number of tissues. Inactivating this gene has a major effect, blocking embryonic development at an early stage.[20] The deficit that leads to embryonic death appears to involve the cells that form the blood vessels. The mutation also induces a significant level of cell death, in particular in nerve cells. The characteristics of the mutant can be explained on the basis of our knowledge of the signals that regulate the formation of the blood vessels or the mechanisms that control programmed cell death. The biologist can thus find a coherent explanation for the anomalies that are found in mutant animals.

Although this explanation makes it possible to interpret the observed phenomena, it does not explain them: explanations that were just as believable could have been proposed if the mutants had shown different characteristics. Indeed, what is surprising about these mutants is not so much the observed defects—most of which seem reasonable, given our present understanding—but rather the fact that they are so limited. Many other tissues and cells in which the Ras protein and the GAP protein that controls it appear to have an important role show no effects whatsoever. This raises the problem of the compensatory mechanisms that mask the role of GAP in these cells and tissues.

This problem is also encountered in explaining the effect of mutations in the genes coding for tyrosine-protein kinases. As explained earlier, one of these enzymes is altered in patients suffering from Williams syndrome. In general, these molecules are involved in various signaling pathways, which interact with the pathway that goes through the Ras protein. The first tyrosine-protein kinase, called Src, was described in a virus which produces a tumor called a sarcoma (hence the name of the protein). This virus has captured a gene from the cellular genome and modified the tyrosine-protein kinase that it encodes. The normal Src protein kinase is probably involved in signaling pathways that enable the rate of cell division to respond to signals from the extracellular environment. If Src is deregulated as in its viral form, it permanently activates cell division, thus rendering the cell cancerous. In trying to dis-

cover how the activity of Src is controlled, researchers found
another tyrosine-protein kinase, Csk (c-src kinase), which in-
hibits Src activity.

These two proteins are expressed in a wide range of tissues. It might be expected that inactivating the gene coding
for the Src protein kinase would have a major effect on cell division, thus blocking development at an early stage. It would
be logical to expect that a mutation in the gene coding for Csk
would have an analogous effect, but of a lesser amplitude—
the destruction of a protein regulator ought to lead to weaker
effects than the destruction of the protein itself.

This is not the case. Inactivating the *src* gene has very minor consequences on development, although the animals do
die a few weeks after birth.[21] Mutant mice suffer from osteopetrosis: their bones are too ossified, to the detriment of the
other normal components of the bone. Destroying the gene
coding for Csk has more important effects that block development—embryos are small, the neural tube does not properly
close, and the circulatory system is absent.[22] In such mutants,
phosphorylation of the Src protein is reduced but it is not absent, showing that there are other protein kinases present in
the cell that can phosphorylate and regulate Src. The results
of the *csk* knockout can be explained as being due to the activation of Src, which has more serious effects than an
underexpression.[23] Another possibility is that the Csk protein
kinase regulates other protein kinases as well as Src and that if
they are deregulated the consequences are serious. However,
all these scenarios are retrospective: nobody predicted the
properties of the mutants. This shows that our understanding
of these regulatory networks is still highly incomplete.

Things do not get any better when we look at the nucleus.
The signal pathway, which began with membrane receptors

and passed via Ras and tyrosine-protein kinases, leads to the activation of a transcription factor called AP1. AP1 activates the transcription of a number of genes, the products of which provoke cell division. The AP1 factor is formed by a combination of a number of Fos and Jun proteins, each subfamily containing three or four proteins. Several genes coding for these proteins have been mutated, and once again the results have been surprising. There is a striking contrast between the phenotype of the mutated animals, which in most cases is almost normal, and the expected effect of the inactivation of genes hitherto considered to be essential for cell division. If the gene coding for the c-Jun protein is mutated, the animal has anomalies in the development of the liver.[24] But the mutation of *c-fos* has a relatively limited effect, altering bone development.[25]

A different kind of surprise was found with the NF-AT transcription factor, which can associate with the Fos and Jun proteins and is involved in the activation of T lymphocytes, the first step of the immune response. The activation of this factor is blocked by the immunosuppressors used in transplant surgery to prevent rejection. The knockout of one of the genes coding for this factor leads to lethal anomalies in the heart valves. The pathology shown by these mutant mice is similar to that observed in certain human congenital heart conditions. This is one more example of a single protein being used for a range of very different physiological functions.

Perhaps the most surprising results so far have been obtained with the mutation of another gene of the Fos/Jun family, which codes for the FosB protein. This experiment was done by two laboratories: for one group the animals were totally normal,[26] whereas the other laboratory found that mutant mice were born normal but many died soon afterward, not because they had anything wrong with them but because their mother, which also bears the mutation, did not take care of them.[27] Researchers studied whether this defect was due to an anomaly in lactation-inducing hormones. No effect could be found, nor was there any defect in the olfactory system, which plays a major role in the attachment of the mother to its offspring. The final conclusion was that the only deficit which could be observed was an absence of maternal behavior. We know that in mice this behavior is controlled by the hypothalamus. When the mother is in the presence of its offspring, the FosB protein is briefly synthesized in this region of its brain; in mutants with inactive copies of the *fosB* gene, of course, this does not take place. The *fosB* gene thus appears to be one of the genetic components that enable mammals to carry out parental care.

Could this result have been predicted? Previous studies had shown that the *c-fos* and *fosB* genes were expressed in the central nervous system and that their expression varied as a function of environmental signals. But that does not mean that anything like such a precise role for the FosB protein had been imagined. This raises the more fundamental question that will be discussed later: what do we mean when we say that a protein is essential for the performance of a given behavior?

Before dealing with this point, however, we need to take the next logical step and examine the genes that are necessary for the realization of all complex behaviors—those involved in learning and memory. This is the subject of the next chapter.

CHAPTER SIX

MOLECULES TO MIND

Even primitive animals such as the sea snail *Aplysia* can remember experiences and change their behavior as a function of these experiences. Neurobiologists have proposed a single mechanism for the many kinds of memory formation that can be described: when two contiguous nerve cells are simultaneously active, the points of contact between them (synapses) become stabilized, thus facilitating the subsequent passage of a nerve impulse between the cells. Memory takes on its many diverse forms through the complexity of the neuronal circuits concerned with this synaptic stabilization.

In addition, memory formation is a complex process characterized by a series of successive phases. A brief phase stabilizes the synapses for a few minutes to a few hours; then a long phase corresponds to a stabilization that may last several years.[1] This long phase requires protein synthesis. The existence of two phases in memory formation has been shown in animals; it is probably also the case in humans, even if, for obvious ethical reasons, it has not been possible to use protein synthesis inhibitors to show that the memory-stabilization phase requires protein synthesis.

In higher organisms, memory stabilization is linked to a particular brain structure called the hippocampus (because of its supposed resemblance to a sea-horse, the Latin name for which is *hippocampus*). Ablation of the hippocampus prevents rats from storing spatial references that enable them to adapt their behavior to their environment. A test that is widely used by researchers to measure the ability of a rat or a mouse to recognize its environment involves putting the animal in a basin full of water that has been made opaque (by adding milk, for example). The basin contains a submerged platform on which the animal can rest without having to swim. Normal animals quickly learn to localize the platform by using cues in the environment. Animals whose hippocampus has been destroyed cannot spatially situate themselves and thus never learn to localize the platform.[2] Humans in whom the hippocampus has been accidentally destroyed cannot form any new memories. They still learn complex gestures but cannot memorize facts or faces. On the other hand, previously acquired memories survive the destruction of the hippocampus.

The hippocampus is not responsible for all memory formation. Nevertheless, the relative simplicity of this structure has made it a model system for studying the mechanisms of memory. Observations of the properties of its neurons have shown exactly the kind of phenomena that would be expected of a neuronal structure implicated in learning: the simultaneous excitation of certain neurons present in this structure leads to the long-term potentiation (LTP) of the synapses linking these neurons.[3] LTP thus corresponds to an elementary form of memory formation. Given that this phenomenon is observed in a brain structure that is known to be essential for the process of memory formation, researchers have been tempted to equate LTP with memory formation. The study of LTP in the hippocampus has thus become a model for the study of the phenomena of memory.

Many studies have been carried out on the mechanisms involved in memory formation and in the appearance of LTP. Different approaches have tended to converge on a small number of proteins, which have been shown by experimental studies to play a key role in memory formation. The first line of research was fairly obvious: given that, in cellular terms, memory phenomena are due to the reinforcement of the synapse following an initial neuronal excitation, the molecular bases of memory processes should be sought in the responses produced by this initial excitation. The excitation of a neuron leads to a number of events: an increase in the level of intracellular calcium, the production of second messengers, the transcription of "early" genes—which are similar to those that are activated in a cell stimulated by a growth factor or a hormone (see Chapter 5). As we have seen, protein kinases are essential constituents of these signaling pathways, and their involvement in memory formation was immediately suspected.

This suspicion was already confirmed by a long series of studies on *Drosophila* begun over 30 years ago by Seymour Benzer and since continued by a large number of other groups, the aim of which is to characterize the genes involved in behavior.[4] Probably the most successful part of this project has been the study of the genes involved in learning, using very simple apparatuses that give flies an electric shock in the presence of a particular odor and then allowing them to avoid the shock-associated odor. A large number of mutations involved in learning and memory have been discovered in this way.[5] Two of these mutations, *dunce* and *rutabaga*, have been particularly intensely studied: they both affect the production of cyclic AMP, a small molecule that is essential for intracellular signaling. Cyclic AMP was discovered in the fifties through its role in controlling the metabolism of glycogen and, more generally, in the metabolic response of cells to hormonal signals. Cyclic AMP activates a special protein kinase called protein kinase A.

Another gene involved in learning in *Drosophila* codes for

a calcium-activated protein kinase, CaM Kinase. When this protein kinase is inhibited, flies show poor learning in a naturalistic learning set-up involving the fly's sexual behavior.[6] A male fly will court both virgin and mated females, despite the fact that, once mated, a female rarely mates again. Normal males rapidly learn to avoid courting unreceptive mated females, whereas males with low CaM Kinase activity continue to court mated females—to no avail.

Encouraged by these results from mutant studies in *Drosophila*, neurobiologists decided to use the knockout technique in mice to see whether analogous genes or signaling pathways were involved in the functioning of the hippocampus and the production of LTP. Many such genes have been mutated and the effects of these mutations described. These relatively straightforward experiments have already shown that a number of genes are involved in LTP and in learning in mice.

In one of the first experiments, the gene coding for N-CAM, a neuron-specific adhesion protein, was knocked out.[7] Mutated animals learn more slowly in the submerged platform test: this macromolecule could be involved in the synaptic reinforcement that is linked to learning.

Given the role of cyclic AMP in learning, as shown by studies of both *Drosophila* and other organisms such as *Aplysia*, it was quite logical to try to alter the signaling pathways that involve cyclic AMP. Various components of this pathway were selected, including one of the enzymes involved in the synthesis of cyclic AMP[8] and a protein, CREB, which, once it has been modified by protein kinase A in response to cyclic AMP, activates the transcription of several genes.[9] Inactivating the two genes corresponding to these proteins did not prevent all learning nor the formation of LTP, but it did reduce the duration of LTP and altered the fixation of memories and behaviors. The cyclic AMP signaling pathway thus appears to be involved in the second phase of memory formation, which depends on protein synthesis and leads to the stabilization of memories and behaviors.

Three other mutations have been shown to affect the formation of LTP: the alteration of Fyn, a tyrosine-protein kinase which, like Src and Csk, is involved in intracellular signaling;[10] a mutation in the gene coding for αCaM Kinase II (the same protein described in *Drosophila*),[11] which leads to reduced spatial learning in mutant mice;[12] and the knockout of the receptor for a neurotransmitter involved in the formation of LTP, called the NMDA receptor. This receptor is located at the neuron membrane where the activating signal arrives, upstream of the protein kinases. The interaction of this receptor and its neurotransmitter leads to an increase in intracellular calcium, which in turn activates the αCaM Kinase II protein. Blocking the NMDA receptor with various specific pharmacological agents had been previously shown to block learning and formation of LTP in the hippocampus.[13] The gene coding for this receptor was not knocked out in all of the organism but only in some cells—the CA1 cells in the hippocampus.[14] This kind of experiment makes it possible to conclude unambiguously that the observed alterations are due only to the perturbed functioning of the gene in the tissues that have been targeted by the researcher. The authors described the precise nature of the deficit in the knockout animals.[15] In normal animals, there is a cognitive map of the environment— each hippocampal neuron is excited when a certain kind of landmark is present in the animal's environment. In mutant animals, this cognitive map is disorganized, less precise, and less stable.

The reverse experiment was also done—researchers increased the level of the NMDA receptor (or more precisely of one of its subunits which is abundant in the young animal).[16] The result was striking: not only did the animals not exhibit any defect, but they also performed better in three different learning tests. However, to describe the experiment as a "genetic enhancement of intelligence"[17] and to suggest that it could be easily reproduced in humans reflects a confusion between memory and intelligence and a profound ignorance of the difficulties involved in transgenesis in humans. Furthermore, it would probably be wise to wait a while before drawing definite conclusions. For example, the first transgenic mice obtained following introduction of the gene coding for growth hormone were twice as big as normal.[18] Only later was it discovered that the fertility of these transgenic animals was profoundly altered. Even if we set aside for the moment the possible application of such transgenesis—and its difficulties—it is clear that these experiments of the NMDA receptor show that this macromolecule occupies a central place in mechanisms of learning.

These studies have brought molecular neurobiologists close to their holy grail—understanding the molecular mechanisms involved in behavior and in cognitive functions such as memory formation, thus linking molecules to mind.[19] They also tell us a number of things about the role of genes in cognitive processes. The first is that the genes involved in learning are not specific to this process. They code for ordinary proteins that are involved in intercellular interactions and intracellular signaling pathways. Even glutamate receptors (the receptor family that includes NMDA) are not specific to the nervous system—similar proteins are found in plants.[20] There are no proteins specific to learning and memory but

rather proteins that, through their function as relays or trans-
mitters, have been harnessed by evolution in the development
of cognitive processes: mutant animals often showed many
other deficits besides problems of memory formation. When
the first genes affecting memory were isolated and characterized in *Drosophila,* there was surprise and disappointment
when it turned out that the proteins coded by these genes
were already well known for their role in metabolism. What
makes a process specific is not the nature of its molecular
components (and thus the genes that code for these components) but the way they are used and assembled in particular
molecular pathways and specific structures.

If, instead of simply knocking out genes coding for these
molecular components, we were to make more subtle changes
—as in the case of the NMDA receptor—they would no
doubt lead to more subtle behavioral changes. It is possible
that over the next few years we will discover that humans have
a number of different alleles coding for, say, one of the subunits of the NMDA receptor and that one of these gene forms
is linked to superior abilities in terms of facial recognition and
learning the environment. But this would not mean that this
gene is *the* gene involved in facial recognition, because, as
shown above, the complex functioning of the hippocampus
(and probably of other parts of the brain), involving the activity of hundreds of genes, is required for this ability.

Biological processes are indeed genetically controlled, but
this does not imply that the gene products are in and of themselves responsible for these processes. They are simply components that participate in these processes. Biological function emerges from the complex organization that spans the
whole scale of life, from molecules up to whole organisms or
even groups of organisms. Complex functions find their origin

and explanation in this hierarchy of structures, not in the simple molecular components that are the direct product of gene expression. It is thus doubly absurd to state, as some biologists have, that the protein kinases described in this chapter are "cognitive" kinases,[21] or to suggest that these kinases are the "engram" or the physical incarnation of memory. The reason is simple: these protein kinases are not specific to memory or cognition, and their action takes on its importance only within the highly organized structures of the central nervous system.

Memory processes are characterized by the fact that organisms can alter their synaptic connections as a function of environmental effects. The genes described here are not involved in the rigid construction of the central nervous system but in its ability to continually modify itself, in its plasticity. "Memory-formation genes" determine the indeterminacy of higher nervous structures.

CHAPTER SEVEN

Genes Controlling Life and Death

This chapter deals with the genes involved in all the stages of life. At the beginning there is development, and at the end there is aging. Although aging leads to cell death, the two processes are very different. Furthermore, cell death is quite unlike the death of an entire organism. As we will see, cell death occurs throughout life, even during development, in most cases for the benefit of the organism. Finally, there is often a link between these processes and cancer: even if cancer can lead to the death of an organism, cancerous cells acquire a kind of immortality and express many of the characteristics of the undifferentiated cells that are present during early development. The questions raised by the study of cancer, as we will see, underline the difficulties that occur when a group of genes is described simply as "the genes for . . . " So it is with the subject of cancer that we will begin.

Genes and Cancer

Between 1975 and 1985 genes were discovered which, when mutated, lead to the formation of cancers. The discovery of these oncogenes represented the first major success of the new technology of genetic engineering.[1]

The idea that tumors could form following the mutation of one or more genes was not new. This hypothesis, originally proposed in the early years of genetics, had attracted a great deal of experimental support in the period immediately preceding the discovery of oncogenes. In particular, Bruce Ames's group had shown that known carcinogenic chemicals were mutagenic agents which could interact with and modify DNA, either spontaneously or after transformation by the organism's detoxification enzymes.

What was surprising, however, was that only a few genes became oncogenes when mutated. Furthermore, they were all linked to the same overall cellular function: like *ras* or *src*, they coded for proteins or enzymes that form the intracellular signaling networks and pathways that enable cells to adapt their rate of division to the needs of the organism.[2] Cancer genes appeared to be the genes involved in the control of cell division, which when mutated provoke the permanent activation of this process. Once the existence of this limited number of genes had been demonstrated, fascinating perspectives in diagnosis and therapy were opened up. From a more fundamental point of view, this important discovery showed that it was possible to isolate groups of genes with specialized functions. All this suggested that there was an oncogene family.

The characterization of genes involved in tumor formation did not lead to the same debates or the same polemics as the characterization of genes involved in human behavior. And, in fact, this absence of any higher stakes, coupled with the large number of studies carried out since the discovery of oncogenes, makes our understanding of their action particularly rich and also reveals the real difficulties that exist in linking a family of genes to a particular function.

The currently accepted model of the nature and function of oncogenes is far more complex than at the beginning of the 1980s.[3] The very category of oncogene has become highly heterogeneous. The first oncogenes to be characterized were "dominant"—they led to cancer even if only one copy of the gene was mutated. A second category was then discovered, anti-oncogenes or tumor-suppressor genes, which when inactivated lead to tumor formation. A third class has since been added, that of "immortalizing" genes. As we shall see, these genes inhibit cell death and favor the appearance of oncogenic mutations. Other genes that favor the growth of blood vessels to nourish the tumor or enhance its ability to disseminate and form metastases could also be added to the list.

As well as this internal complexity in the category of oncogenes, there is also the problem of the ever-increasing number of genes that are thought to be involved. It could be argued that this is simply a reflection of our growing understanding of the signaling pathways in which oncogene products intervene. The total number of oncogenes should therefore eventually stabilize and equal the number of genes coding for the components of these pathways. Things don't work that way, however. It rapidly became apparent that some genes involved in the construction of these networks were more likely than others to produce oncogenes when mutated. On the other hand, some genes that code for key components of these pathways never become oncogenes.

We encountered these kind of difficulties earlier in some cases of genetic diseases and the effects of gene knockouts. The regulatory networks involved are highly complex, with parallel pathways and even points at which the pathways intersect. This makes it difficult to predict the effect of perturbing even a single component. There is no straightforward relation between the overall behavior of a network and the alteration of any one of its components.

Although these problems make it very difficult to define and identify the members of the oncogene family and to develop diagnostic and therapeutic applications, they should not make us forget the main finding: the genes identified by these methods do indeed play a major role in tumor formation. Introducing a dominant oncogene and an immortalizing gene, while at the same time inhibiting two of the main tumor suppressor genes, is sufficient *in vitro* to transform human cells that were completely normal into cells that have the properties of cancerous cells.[4]

It should be noted that many oncogenes also belong to the family of developmental genes. In itself, this is not surprising: like cancer cells, cells in the embryo divide very rapidly, and cancer cells often express embryonic characteristics. Indeed, many biologists have underlined the inverse relation between cancer and cell differentiation.[5]

This difficulty in providing a strict definition of gene families, with precise and distinct functions, does not abolish the power of genes. It simply gives it a particular coloration.

GENES AND DEVELOPMENT

Embryologists found it very difficult to accept that genes control development. The reason was that many of the fundamental processes of development, in particular the establishment of the body plan of the organism, are in many species based on properties of the maternal cytoplasm and not on the activity of the nucleus.

Today, these quarrels between geneticists and embryologists are obsolete. We know that genes play an essential role in development, even in the early stages. In some species, maternal genes synthesize components that are contained in the egg and which organize the first stages of development. In other species, such as mice and humans, the embryo's own genes organize each step in development by making it possible for the embryo to synthesize certain molecular products. Thus, while development is under genetic control—a banal idea because nothing escapes genes—genetic control is rarely as straightforward as it is generally presented.

The most original discovery in the field was the demonstration that a certain number of genes play a specialized role in development. This idea is relatively recent in terms of the age of genetics, and its origins can be attributed to François Jacob and Jacques Monod.[6] At the end of the 1950s, they distinguished two types of genes—structural genes which synthesize various protein components that ensure cellular structure and function, and regulator genes which—through the synthesis of regulatory proteins—regulate the expression of structural genes, either directly or indirectly (by controlling the function of other regulatory genes). One of the characteristics of complex life forms like humans is that different cell types express different proteins. For the early molecular biologists, explaining development required explaining the progressive appearance of this regulation of gene expression. Development was thus a question of regulatory genes.

Mary-Claire King and Allan Wilson compared the genomes of humans and chimpanzees, two organisms that appear to be very different.[7] However, if we analyze the proteins that form them, for example blood proteins, they are remarkably similar. On the other hand, two species of fruit fly that cannot be distinguished by the naked eye may be formed of proteins with relatively different structures. Genetic differences thus do not necessarily tell us anything about morphological differences. For King and Wilson, morphological differences result from differences in the construction of organisms and thus involve only regulatory genes rather than structural genes. A simple difference in the time a particular

structure develops can have major developmental repercus-
sions and may modify the final form of the organism, without
changing the nature of its constituent proteins in any way.[8]
The evolution of organisms would thus reflect modifications in
genes that regulate development.[9]

The idea that development is controlled by the action of a
handful of genes, although seductive, was accepted only with
difficulty. First, embryologists refused to concede that the
complexity of development could be reduced to the action of
mere genes. Second, orthodox evolutionists preferred to con-
ceive of evolution in terms of the accumulation and selection
of a large number of genetic modifications, each of which had
a very small effect. For many of these biologists, the concept
of developmental genes conjured up the specter of Richard
Goldschmidt, the German geneticist who had opposed both
neo-Darwinism and the concept of individual genes.
Goldschmidt had argued that evolution could not be explained by the accumulation of micromutations but only by
macromutations—sudden jumps that profoundly modify the
structure and function of the organism, leading to the appearance of what he termed "hopeful monsters."[10] Even more seri-
ously, in the 1960s the existence of developmental genes was
based on no experimental data whatsoever, either genetic or
molecular.

Demonstrating the existence of developmental genes was
the work of a group of *Drosophila* geneticists: Antonio Garcia-
Bellido, Peter Lawrence, Ed Lewis, Walter Gehring, Eric
Wieschaus, and Christiane Nüsslein-Volhard.[11] This work was
so important that Lewis, Wieschaus, and Nüsslein-Volhard
were awarded the Nobel prize in 1995. They showed that
genes were indeed essential for the first stages of development in this organism. But only after developmental genes

had been cloned and characterized at the beginning of the 1980s, using the new tools of genetic engineering, were they universally accepted. The key result that convinced even die-hard skeptics of the importance of these genes was the discovery that they were remarkably well conserved over evolution: these genes are essential components of development in organisms that differ in their final structure, and even in the mechanisms that put this structure into place.[12] This conservation is so astonishing that Jonathan Slack has even proposed that their presence in an organism is sufficient to define it as an animal.[13]

What are these developmental genes? As Jacob and Monod suspected, many of them are indeed regulatory genes, the products of which control the expression of other genes by acting as transcription factors. The most famous of these genes are undoubtedly the *Hox* genes, called homeotic genes because when mutated they lead to the replacement of one part of the animal's body by another. Grouped together in the genome in one or a few gene complexes, depending on the species, these genes are successively expressed during development and specify structures along the antero-posterior axis of the animal. Other genes, such as the *otx* genes, are essential for the formation of the head. There are also regulatory genes that are involved in later stages of development, not in organizing the embryo's body plan but in forming differentiated tissues such as muscle (for example, the *myoD* gene).

However, regulatory genes are not the only developmental genes. This category also includes a very large number of genes that code for proteins involved in intercellular signaling, for membrane receptors, or for enzymes that intervene in these signaling pathways. These communication genes are important both for the development of the general structures of the embryo and for the precise construction of organs. The discovery that developmental genes included genes coding for components of the inter- and intracellular signaling pathways was vital for convincing molecular biologists, or at least the most reductionist among them, that development could not be described only at the genetic level. Understanding development requires the study and description of the hierarchical structural organization of organisms, particularly at the cellular level. Developmental genes do not directly construct the organism; rather, they provide cells with the relevant properties that enable them to interact in order to construct the organism. Thus, although Jacob and Monod's models played an important role in the birth of the concept of the developmental gene, this concept now has a very different form, corresponding to the new molecular vision of the 1990s.

As well as the structural conservation of these developmental genes, there is also a functional conservation: a *Drosophila* gene of a given type can be replaced by its mouse equivalent, and vice versa. Moreover, evolution has not conserved isolated genes but rather pathways and entire networks of developmental genes—functional modules.[14] However, the number of developmental genes in vertebrates is far greater than in insects—this is another example of the repeated duplication of the genome that has taken place over evolution. In the case of development, this redundancy might be used to ensure the efficacy of this highly complex process. Gene duplication might channel development, making it more resistant to environmental perturbations and thus more secure.[15]

Such a possibility was proposed more than 50 years ago by Conrad Waddington and Ivan Schmalhausen.[16] One of the best examples of this functional redundancy of developmental genes is that of the genes responsible for the formation of muscle cells, in particular the *myoD* gene and the *myf5* gene. Knockouts of each of these genes have a very limited effect. On the other hand, the simultaneous inactivation of both genes prevents all formation of skeletal muscles.[17]

The conservation of developmental genes might reflect the conservation of developmental mechanisms. If similar genes are involved in determining the anteroposterior and dorsoventral axes of insects and vertebrates, this is because these organisms use the same molecular mechanisms to construct themselves as did their long-dead common ancestor. The resemblance of genes is a mark of a homology in the construction of organisms. Nearly two hundred years ago, two French scientists—Geoffroy Saint-Hilaire and Georges Cuvier—argued over whether the body plans of insects and vertebrates were related. Saint-Hilaire argued they were; Cuvier argued they were not. We now know that Saint-Hilaire was right.[18]

However, we should avoid jumping to conclusions: using the same genes in development does not necessarily indicate the presence of a homology.[19] Within a given organism, developmental genes are used over and over again—"recycled," to use the term adopted by some scientists—for making different structures. This is the case for the genes in the homeotic complexes which, as well as determining the antero-posterior axis of animals, are also employed for making limbs, the urogenital apparatus, and the digestive tract. Recently, it has even been shown that one of these genes is involved in the development of hair.[20] Thus the fact that two different organisms use the same developmental genes to build the same structural element does not imply that the common ancestor of these two organisms also used the gene for that purpose. The genes were present in the common ancestor, and their use in the same developmental step might have been a parallel and independent invention of the two species, an example of convergent evolution.

What exactly do developmental genes do? The homeobox genes were the first developmental genes to be described, and they are still the best known. In the fruit fly, these genes determine the nature of the fly's segments. In the mouse, the results of the first knockout experiments were unclear. Only some mutations had homeotic effects, transforming one part of the body into another.[21] Other mutations had more complex effects that were difficult to interpret. The models proposed to explain the action of these genes in mammals are still vague: rather than having a direct influence on the nature of the organs, it is thought that these genes have an indirect effect on morphology, via a regulation of the rate of cell division during development.[22] In fact, homeobox genes can be activated during tumor formation.[23]

The difficulty in interpreting the effect of knockouts of such genes flows in part from the fact that the genes have been duplicated during evolution. Whereas the fruit fly has only one copy of a given gene, the mouse has four, with partially redundant functions. The other problem relates to the phenomenon of phenotypic convergence that has already been referred to a number of times. Different mutations, in different homeotic genes, can produce the same phenotypic change. For example, the presence of a fourteenth rib in a mouse called Adam (in reference to the Bible) may be caused by the alteration of several different homeotic genes. Forming the right number of ribs is a weak link in development that can be revealed (rather than induced) by several different mutations. These results suggest that there are constraints during morphogenesis that are merely modulated by the action of developmental genes.[24]

Other experiments, however, show that some developmental genes are indeed, as Walter Gehring has called them, "master control genes."[25] The most striking example is that of *pax-6*, a gene involved in eye development in fruit flies and in humans.[26] If this gene is expressed in parts of the fly's body where it is not normally expressed, such as on its wings or legs, extra eyes appear in these parts of the fly's anatomy.[27] This somewhat grotesque experiment is utterly astonishing: the construction of a fly's eye involves more than 2,000 proteins, and thus the coordinated expression of over 2,000 genes. The structure and function of the *pax-6* gene has been conserved over evolution. A similar result has been obtained in amphibians.[28] Overexpressing the mouse version of the gene in the fly leads to the growth of extra fruit fly eyes: the master control gene does not determine the precise nature of the structure that is produced.

Somewhat paradoxically, the *pax-6* gene, which seems to be so specific to eye development, is also expressed in many embryonic structures, in particular in various sensory structures. Furthermore, in the mouse, the gene is necessary for the development of the pancreatic cells that synthesize glucagon, a hormone that balances the action of insulin.[29] Several other genes show similar powerful effects as *pax-5*, intervening in a virtually identical fashion in eye formation, at least in *Drosophila*.[30]

Another element of complexity needs to be added to the definition of the role of genes in development. There are many examples where the structure of these genes has been conserved over evolution but where they nevertheless have very different functions in different organisms. For example, the same genes, present in the same signaling pathway, are involved (1) in the construction of photoreceptor cells in the fly's eye, (2) in the construction of the nematode's vulva, and (3) in the control of cell division in mammals. This shows that a developmental gene in one organism may not be a developmental gene in another organism but may simply participate in the life of the cells.

In the earlier discussion of genome sequencing, I pointed out that one of the striking results obtained by the analysis of the yeast genome was the abundance of genes coding for proteins involved in transport and communication, that is, of genes that enable the yeast to adapt to changes in its environment. In building ever-more complex multicellular organisms, nature has used—and sometimes perfected—these signaling pathways, thus making it possible for cells to exchange information during development. Some of the genes that code for components of these pathways have become developmental genes. Others have kept their original functions, while participating in development.

Like the family of cancer genes or memory genes—or, as we will see, behavior genes—the family of developmental genes is an open family, with a very free lifestyle; the term "community" would probably be more accurate. Of course, this community only exists in the minds of biologists—another indication that our concepts are inadequate and that reality constructs and organizes itself according to principles that are not ours and which remain fundamentally mysterious.

Long before these observations were made, François Jacob proposed the metaphor of "tinkering" to account for the surprising results of the early molecular studies of higher organisms which showed that, over evolution, the same genes, or the same parts of genes, had been used to carry out very different functions.[31] Unlike an engineer, evolution does not make plans of the organisms that are to appear. Rather, like

someone tinkering, it shapes organisms with what it has to
hand, using the same tools to accomplish the same objective
or for different objectives. This metaphor explains both the
conservation of the molecular components of life and their
varied uses. As we have seen, it applies particularly well to developmental genes.[32] Like evolution, a tinkerer does not respect the specificity of the objects to hand but simply uses
them as a function of their properties.

This metaphor, while it is undoubtedly seductive, has
weaknesses. The first is the clear distinction it makes between
the work of an engineer and that of a tinkerer. Such a distinction is not valid: like tinkerers, engineers also build with what
they have to hand, with what they know. Although engineers
may have more leeway than tinkerers, they are still limited by
the materials available, by their knowledge, and perhaps most
of all by the past, by what has previously been created. These
theoretical and practical limits effectively transform the engineer into a super-tinkerer.

However, the main weakness of the metaphor of tinkering
to describe the work of evolution is that it masks the essential
difference between any human labor and evolution. The engineer and the tinkerer both have a clear vision of what they
want to build, whereas evolution is blind. If we wish to conserve this metaphor, we might think of the tinkerer of evolution as a modern sculptor who gathers and arranges various
objects to see what they look like together. Even this analogy
probably does not hold, because this sculptor already has a
reasonable idea of what he or she wants to create, even if the
final details will depend in large part on the haphazard nature
of the objects at hand.

Despite these weaknesses, the tinkering metaphor under-
lines the key problem faced by developmental biologists. If

evolution, as it has taken place on our planet, has been the result of tinkering with a kit of developmental genes established shortly after the appearance of multicellular life, what are the constraints that this gene-kit has imposed on the construction of organisms? What can it tell us about evolution?

Most population geneticists and neo-Darwinians would probably reply "Nothing." The rules fixed by these genes are sufficiently supple for the final result to depend more on the process itself than on the rules that govern it. Indeed, the study of developmental genes may turn out to be fruitful in terms of describing the evolution of life but not in terms of understanding it. We cannot be sure that the structure and properties of the products of these genes contain something rich and strange that provided organic evolution with some of its key characteristics. In itself, their conservation and repeated reuse does not prove their importance—it may simply be the result of a principle of parsimony that guides the functioning of all organisms.

The only argument that can be put forward is retrospective. Twenty years ago, no biologist, and in particular no evolutionary biologist, would have dreamed that it would be possible to study the evolution of life by studying a small family of genes that intervene in development. Such an idea would probably have been profoundly shocking. Today, all biologists accept the idea that the study of such genes is essential to describing evolution. And tomorrow?

GENES AND AGING

Up until a few years ago, the genetics of aging was largely unknown territory. There are many reasons for this apparent lack of interest: aging is a very difficult area of research in which the subject is hard to define. Furthermore, it is dogged by an unpleasant scientific past in which some brilliant minds went astray. Many biologists have become interested in the problem of aging at precisely the point at which old age weakened their abilities while at the same time increasing their motivation—perhaps too much. In other words, it is the kind of subject that scientists learn to avoid in order not to get bogged down, preferring doable problems to metaphysical quagmires.

Nevertheless, a number of the characteristics of aging make it a fascinating subject, particularly in relation to the role of genes. Aging can be studied at all levels of organic organization—molecular, cellular, and organismal.[33] The relation between these different descriptions is difficult to discern. Each phase in the history of science has emphasized one particular level of organization and has located an explanation of aging at this level. Contemporary science is no exception: current research aims at finding the molecular bases of aging. And yet any satisfactory description and explanation of aging will have to include data from several different levels of organization and will not be reducible to any single factor. More than in any other area of research, it is clear that reductionist or holistic approaches to aging—bottom-up or top-down explanations—are too intellectually impoverished to fully explain the phenomenon.

All research on the subject shows that aging is a highly complex process associated with two related but distinct processes—illness and death. Illness often accompanies aging, and many diseases become more frequent with age. But aging is different from illness: it is possible to age without becoming ill, despite showing all the classic signs of growing old. Aging is also different from death: it is possible to die without aging—

some people die young, although the risk of dying grows with age. Some changes characteristic of aging (perhaps all of them) also increase the probability of dying from illness. It is thus difficult to define aging, the process being closely linked to both life and death. There is little hope of characterizing the genes involved when the process itself cannot be precisely defined! (This difficulty is also found when it comes to defining development or intelligence.)

Research on aging is interesting for another reason. As we will see, many genetic changes lead to changes in the aging process. However, for reasons that will be explained, there are probably no "aging genes"—genes, that is, whose role is to program aging. What makes genetic studies of aging so important is that they show how genes can intervene in a process that they do not control.

The main reason, however, for dealing with this topic is that a series of genetic studies have recently been carried out on a number of animal models (in particular the nematode and the fruit fly), as well as some initial investigations of humans. Although these data are incomplete and difficult to interpret, they clearly show the place of genes in aging. Given that until recently relatively few studies had been carried out on aging, it is worth giving a brief summary of the main results that have been found and the theories that have been put forward to explain them.

Life spans vary enormously between species, and aging takes many different forms. Despite this diversity, some phenomena are relatively general: most organisms age, even simple organisms like paramecia or yeast. In yeast, aging is expressed by the inability to produce daughter cells; in this species, therefore, aging is not measured in years but in cycles of replication. In more complex species, the maximum life span is both remarkably constant within a species (for a given environment) and highly variable between species: some insects live only a few days, whereas some animals and plants can live hundreds of years. This tight control of the life span within a species, added to a strong variability in life spans across species, shows that life span and the aging that accompanies it and leads to death are both closely dependent on those factors that ensure the stability of each species' characters over the generations: genes.

At the level of the organism, aging is accompanied by a change in the functioning of all the organs. The most affected organs are the brain and the sense organs. Less obviously, but just as importantly, the immune system undergoes a profound change that weakens its response to external stress and sometimes even leads to its turning against the organism itself, in the form of autoimmune diseases.

Many alterations also take place at the molecular level as organisms age. For instance, there are changes in the proteins of the extracellular matrix that links and surrounds cells. These changes lead to the loss of tissue elasticity that is characteristic of aging. Less obvious changes have recently been revealed by molecular studies. Many proteins are modified either during or after synthesis, to the extent that they can no longer carry out their normal function. These changes can even lead to the formation of toxic breakdown products. Age-related modifications can also affect the genetic material, or the expression of this material in proteins. It has been shown that aging is often accompanied by changes in the DNA molecules contained in the mitochondria that are responsible for the synthesis of some constituent elements of this organelle. The cell nucleus and the chromosomes are also modified: in some species the extremities of the chromosomes—the telomeres—gradually get shorter, which can cause cell division to cease eventually. The expression of many genes is also altered. This result was already known by using "classical" techniques, but it has been confirmed by functional postgenomic approaches,[34] although it is not yet understood what causes this. This kind of change (mainly?) affects those genes that are involved in the defense of the organism against external stress—toxic components, sudden changes in light and temperature, as well as genes involved in cell division.

Midway between physiological and molecular changes, aging also leads to cellular changes. The most obvious sign of cellular aging is the gradual loss of the ability of cells to divide.[35] One of its clearest expressions in old people is that wounds tend to take longer to heal.

All levels of organic organization thus express the phenomena of aging. At each level aging can be traced by observing many different parameters. Scientists have always tried to find some kind of order in these data, seeking to introduce a hierarchy between those that relate to the same level of organization, to link those that relate to different levels, and thus to define the most fundamental observations, with the intention of finding the causes of aging. Many scientists have tried in the past; many are trying today.

The wide variety of models that have been proposed should lead us to be prudent—they tend to reflect the knowledge and fashions of each period rather than the hidden order of things. At the end of the nineteenth century, aging was seen as being due to some glandular deficit; today, chromosome defects or gene mutations are thought to be the cause. Many studies are currently focusing on the DNA present in the mitochondria, the nucleoli where the ribosomal RNAs are synthesized and matured, or the telomeres (the ends of the chromosomes). Telomerase is an enzyme that makes the telomeres longer; introducing the gene coding for this enzyme into cells that lack it increases the ability of these cells to divide.[36] However, this does not necessarily tell us anything about the consequences of such an activity in a living animal. Would there be an increase in the life span of the organism? Or would the introduction of the enzyme have bad effects, favoring the multiplication of abnormal, cancerous cells which would otherwise have died?

Despite these caveats, many authors have correctly pointed out that the most obvious deficit of age, at both the organismic and the molecular levels, is the inability to respond to the external milieu. In terms of the whole organism, the sensory organs, the cognitive structures, and the immune system are most affected. This observation may simply reflect the fact that these systems are those that are most frequently activated during the organism's life and are the most important for its survival.

It would be imprudent to go any further in describing a hierarchy of changes associated with age. Why do such changes take place? Are they modifications in gene products—proteins—or are they the direct result of changes in the genes themselves, thus producing alterations to the proteins they code for? In answering these questions, the partisans of an epigenetic explanation have fought it out with the partisans of a genetic explanation. In itself, there is nothing new about this. It is the reflection of the long-standing debate between those who consider that genes play an essential role in the organization of organisms and hereditary phenomena and those who attribute a central role to proteins and to the cytoplasm.

The same clash of ideas took place over the origin of cancer. In that case, however, the debate has recently been clearly resolved in favor of the genetic origin of cancer, in that cancers are always the result of modifications of genes, even if these changes may be induced by environmental factors such as tobacco or asbestos. The debate over aging may come to the same conclusion, but we cannot be certain. Whereas tumor-formation implies that the transformed cell multiplies and transmits the transformed character to its daughter cells, the same does not hold for the aging process, which affects the cells and not their offspring. The nervous system ages—indeed, it is one of the systems that is most sensitive to aging—whereas most of the cells that make it up, at least of the neurons, do not divide and multiply at all. However, if one of the changes associated with aging were to be coded in the genes of one of the cells that continually multiplies, this character would be automatically amplified. Genetic and epigenetic aging are equally probable, although the former could have more serious consequences.

By "genetic aging" I mean that aging is accompanied by changes in certain genes that alter their function, and not that there are particular genes involved in aging that are more "responsible" for this process than other genes. A number of laboratories, however, are actively searching for such genes "for" aging. Over the last few years, various experimental approaches have made it possible to study directly the genetic control of aging.[37]

A few studies have been carried out on human beings. It is said—although in practice it is difficult to verify—that there are families of centenarians. If this were to be true, such extraordinary longevity would not necessarily indicate the involvement of genetic factors. It is at least as probable (if not more so) that some common lifestyle could explain the phenomenon, especially given that, in the absence of unusual circumstances such as isolation of the population and high levels of inbreeding, the character would rapidly disappear. As seen earlier, the classic way of showing that a given trait is genetically determined is to correlate the transmission of the trait with a certain number of genetic markers. This type of study, currently widely employed in investigations of the susceptibility to various diseases, is all the more effective if the number of genes involved in the control of the trait is relatively limited.

Longevity, however, presents a major problem for this kind of study. The character can only be determined after death. Families of centenarians exist only on paper, in registers of births and deaths. Geneticists can, of course, study individuals who are already older than average and are therefore long-lived, even if the age of their demise remains unknown. The aim is to see whether such individuals have a higher frequency of certain genetic markers than average members of the population. A statistically significant bias in the frequency of a particular form of a gene necessarily implies that this gene, or a gene close to it, is linked to the observed character. However, this method tells us nothing about the nature of the link between the gene and the character under study. Its most important weakness is that, by its very nature—or at least given the current state of the art—it is biased by previous knowledge: only genes that are suspected to play a role in the character will be investigated further. Fortunately, this limitation is merely temporary. Given the progress of the various genome sequencing programs and the development of DNA microarray technology, it will probably be both possible and practical to look for particular forms of an ever-larger number of genes, without having to restrict the search to those genes that are thought *a priori* to play a role in the function under study.

Such studies have shown that in humans, certain forms of two genes coding for serum proteins are more frequent in centenarians. One of these genes codes for apolipoprotein E (ApoE), which is involved in lipid transport in the blood. The other gene codes for an enzyme (ACE) that modifies angiotensin, a protein involved in the regulation of blood pressure. Both genes code for proteins potentially involved in pathological symptoms of aging—atherosclerosis and high blood pressure. However, the results are not easy to interpret: the particular form of ApoE that is more frequent among centenarians is associated with a lower risk of developing Alzheimer's disease,[38] but, paradoxically, with an excessively high concentration of lipids in the blood. And the form of ACE that is more frequent among the long-lived has been shown to be associated with a *higher* risk of heart attack. Further studies will either confirm or disprove these results and will almost certainly add to the number of genes associated with longevity. Once again, however, it should be remembered that an association between a particular allele and a given phenotype—in this case aging—does not necessarily imply a simple causal relationship between gene and phenotype.[39]

Another approach used in studies of humans is to characterize the genes involved in certain rare forms of genetic diseases that lead to premature aging. Down syndrome, which is caused by having three copies of chromosome 21 instead of two, leads to the early appearance of a number of diseases that are normally associated with old age, suggesting that one or more genes on chromosome 21 might be involved in aging. However, very many genes occupy this chromosome, and characterizing the genes potentially involved will be very difficult.

Werner syndrome, which is caused by a simple mutation,

at first appeared to be more promising. Patients with this disease die before the age of 50, looking very old. At 20 years old, their skin loses its suppleness, their hair goes gray, and they get cataracts. They develop diabetes, cancer, and heart problems far earlier than unaffected subjects. The gene involved in Werner syndrome has recently been sequenced: it codes for a helicase, an enzyme involved in the uncoiling of the DNA double helix.[40] The double helix has to be uncoiled for DNA duplication and damage repair to take place. This lays the ground for a satisfyingly simple story. In individuals suffering from Werner syndrome, damage to the genome caused by an incorrect duplication process, by toxic chemical agents—produced by the environment or by the organism itself—or by various types of radiation is not repaired or is repaired inadequately. This in turn leads to an accumulation of cancer-inducing mutations and the various functional deficits associated with aging.

This hypothesis both satisfies and disappoints. It is satisfying because it explains how a single gene can have such major effects on the organism; it is disappointing because it does not in fact tell us much about the aging process, apart from the fact that it is linked to a number of cellular malfunctions that Werner syndrome merely imitates and accelerates. "Imitates" because sufferers of the syndrome show an aging process that is similar, but not identical, to that of normal subjects. More precisely, they do not show some pathologies that are often associated with aging, such as Alzheimer's disease or high blood pressure. They do indeed have a higher rate of cancers—apart from a few special types, cancer is a disease of old age—but these cancers are different, affecting the internal tissues, whereas most human cancers affect the epithelia—tissues in contact with the exterior.

The third approach, impossible to pursue in humans, is to create, through artificial selection, strains of animals with an altered longevity, then to characterize the genes that underlie this change. In every species in which this experiment has been attempted, it has been possible to alter life span. However, longevity is never increased substantially. Two organisms—the fruit fly and the nematode—have been particularly studied, because of their short life span (which makes the selection experiment relatively easy) and the fact that their genetic maps are already well-known (which makes the second half of the experiment much easier).[41] For various reasons linked to the mechanisms of reproduction of these two organisms, there have been more studies on the nematode and the results have been more informative than those on *Drosophila.* However, the two sets of experiments are not contradictory, with the fruit fly studies tending to confirm the nematode findings.

In the nematode, several genes have been cloned and characterized which, when mutated, increase the life span of the animal. The first group of genes was already known to be involved in a particular phase of the worm's life cycle. If the population density of worms is too high, or if food resources are insufficient, larval nematodes can transform themselves into a dormant state in which their activity and metabolic rate are much lowered. This stage is known as the dauer. The genes involved in the formation of the dauer are active only during larval life. However, variations in these genes prolong the life of the nematode at the adult stage. Given that the formation of the dauer corresponds to a slowing down of the worm's metabolic activity, it is reasonable to imagine that the genes involved in dauer formation might, even at the adult stage, slow down metabolism. One of the genes involved in dauer formation and the lengthening of life span was sequenced and turned out to be analogous to a gene in higher organisms that codes for the insulin receptor, thus confirming the link between metabolism and life span.[42]

The second kind of mutation that has been described—*clk-1*—also supports this hypothesis. Mutant worms not only live longer, they also develop slower. Adult mutants move, eat, and defecate more slowly than wild-type worms. It is as though the mutants live longer but less intensely. Sequencing the gene has shown that it is homologous to a gene that enables yeast to metabolize its food.[43]

The existence of a link between life span, metabolism, and feeding had long been suspected. A low-calorie diet prolongs the life of mice and rats, and, if preliminary data are to be believed, of monkeys as well.[44] On the island of Okinawa, where inhabitants allegedly have a well-balanced low-calorie diet, there are 40 times as many centenarians as on the other islands of Japan. A biochemical explanation of the benefits of reduced calorie intake has been proposed: all food consumption provokes the formation within cells of oxygen-derived chemicals such as superoxide ions which are highly reactive and thus very dangerous for the organism. These active forms of oxygen are a kind of inevitable by-product of food combustion. Slowing down the metabolism, the rhythm of life, would thus lead to a reduction in the production of these toxic compounds and of the damage they cause to the organism, which leads to aging. These nematode mutations would thus have the opposite effect to the mutation responsible for Werner syndrome in humans.

A third type of mutation that has been isolated in the nematode and shown to prolong the life span of this organism is *Age-1*. The gene codes for a component in a signaling cascade and is probably linked to dauer formation. Mutant animals show increased resistance to a variety of forms of stress—oxidizing agents, heat, or ultra-violet light. They overexpress an enzyme, superoxide dismutase, the role of which is to reduce the concentration of these superoxides. The link between life span and resistance to stress is so strong that nematodes have only to be heated slightly—which renders them more resistant to subsequent stress—to increase their life span.[45]

Studies of *Drosophila* have confirmed these findings: flies selected for greater longevity are more resistant to superoxide ions. The long-lived *Drosophila* mutant *methuselah* is also resistant to high temperature and oxidative stress.[46] Furthermore, inactivating the gene for superoxide dismutase shortens life span in the fruit fly, whereas overexpressing it increases life span.[47] In mice, inactivation of the gene coding for a protein repressing the oxidative stress response leads to an increased life span in this organism.[48]

All these results suggest a very simple view of aging as being due to the damage induced by the by-products of oxidative metabolism.[49] To live longer, all we would have to do is to slow down our metabolism and our rhythm of life or defend ourselves more effectively against the toxic agents produced by this metabolism. According to this view, life span is the result of molecular alterations that are inherent in the very functioning of life and the effectiveness of the means of defense at the disposal of a given organism.[50]

But there is still a debate: more elaborate molecular models of aging involving chromatin silencing have been proposed.[51] In addition, we must not forget that aging is not simply a molecular or even a cellular phenomenon. By making nematodes with a mosaic pattern of expression of the *daf-2* gene in its tissues, Cynthia Kenyon's group has shown that increasing organismal longevity can occur even if the *daf-2* gene activity is modified in only a proportion of the worm's cells.[52] Life span is regulated by signals coming from the reproductive system[53] as well as by sensory perception.[54] This is an elegant reminder that the aging of an organism is not the sum of the aging of its component cells but the "integration" of this aging through intercellular signaling within the environment of the organism.

None of these studies was able to isolate genes that directly control aging. As Peter Medawar initially suggested from a theoretical point of view, a genetic program that controls aging is nonsense. Life span and aging are not processes that can be subject to natural selection, or at least not directly. Nature selects individuals according to their reproductive capacity. Once the reproductive phase is over or has slowed down, the organism escapes the clutches of natural selection. Genetic changes that are expressed at an advanced age and that increase the probability of death would almost certainly not be eliminated by natural selection, even in animal societies (including human societies) where postreproductive individuals contribute to the fitness of reproductives. This does not mean that aging is entirely independent of natural selection—G. C. Williams suggested that the genetic forms that are selected even to ensure effective reproduction could have negative effects on a species' life span.[55] Aging could be the price that has to be paid for effective reproduction.

Many genes are probably involved in aging, even if that does not imply that aging is subject to a genetic program or that there are genes "for" aging. It could be argued that the case of aging is different from all the other examples dealt with in this book, because this process, by its very nature, is not subject to optimization by natural selection. My view is the opposite: far from being an exception, aging represents a highly general case that highlights the real role of genes. The particular and unusual cases are rather those that involve a character that has been subjected to such a high degree of selection that only a few genes seem to be involved in the function. The complexity of biological processes, the number of proteins involved and their multiple roles, the redundancy and functional compensation that can occur at each level of organization are all phenomena that usually buffer the action of natural selection. In most biological situations, natural selection has molded gene function without, however, creating a straightforward link between a few genes and the selected character.

What does this survey of genes and aging tell us about the possibility of intervening against the effects of aging? At the very least, we know that it will be difficult. It might be possible to delay aging by modifying genes that code for proteins involved in metabolism or in stress responses. Activating telomerase might stimulate the regenerative power of tissues without increasing the frequency of tumors. The premature aging of Werner syndrome is caused by a defective helicase; mutating the homologous yeast gene also leads to premature aging.[56] The role of this enzyme thus seems to be very important. By improving its functioning in humans, one day we may be able to slow down the aging of the skin or the appearance of a cataract. However, the action of this helicase does not appear to control the formation of most cancers or the development of Alzheimer's disease. This implies that modifying this helicase would not alter the age-dependent increase in the frequency of these diseases. Any genetic intervention of this kind would probably be akin to being marooned in a leaky boat. Whatever you do—patch this plank, plug that hole—you are merely putting off the inevitable. As each leak is repaired a new one will appear, requiring more effort to repair it. The youth gene, or rather the antiaging gene, is a mirage.

At this point, it is worth repeating a story from the Darwinian psychologist Nicholas Humphrey, which has been retold by Richard Dawkins.[57] Henry Ford, founder of the Ford car company, "commissioned a survey of the car scrapyards of America to find out if there were parts of the Model T Ford which never failed. His inspectors came back with reports of almost every kind of breakdown: axles, brakes, pistons—all were liable to go wrong. But they drew attention to one notable exception, the kingpins of the scrapped cars invariably had years of life left in them. With ruthless logic Ford concluded that the kingpins on the Model T were too good for their job and ordered that in future they should be made to an inferior specification." Dawkins recounts this anecdote to illustrate the fact that, like Ford, natural selection is guided by the principle of parsimony. This story can be turned the other way round: if we are like the Model T, how many parts would we have to change, how many genes would we have to replace, in order to avoid the breakdowns of aging?

GENES AND DEATH

In the previous section a distinction was made between aging and death. Although aging weakens the organism, death is not its natural consequence but is always the result of some kind of accident.

What is true for the organism, however, is not true for the cell. Cell death is part of the normal functioning and development of all organisms, long before the whole organism dies. From studies that began in the mid-nineteenth century, biolo-

gists have discovered that cell death accompanies both metamorphosis in insects and amphibians and the development of
the nervous system in vertebrates.[58] It is only relatively recently that this "normal" cell death has been distinguished
from accidental cell death.[59] The cell's walk down death row
generally follows the same steps: mitochondrial functioning is
disturbed, the nucleus condenses and breaks up, DNA is cut
into short sequences, and the cell implodes into several fragments, each of which is soon consumed by neighboring cells
or by macrophages. This kind of cell death, called apoptosis
after the natural fall of leaves in autumn, is silent: it does not
provoke any inflammatory reaction. And it is so rapid that often it is not even noticed. It is also programmed: the cell that
is about to die has to express a certain number of genes; thus
the program is internal rather than external. However, it
would be wrong to describe this phenomenon as suicide—
"euthanasia" would be a better term, because the surrounding
environment begins the process, either by not providing
something that is essential for cell survival or by providing a
ligand that fixes on a cell receptor and, by a "kiss of death," ini-
tiates the self-destruction program.

Geneticists have recently characterized the genes involved
in apoptosis. Once again, the nematode *C. elegans* has been
the model. The development of this organism is strictly programmed, and the number of cells in the adult (959, excluding
the cells in the germ line) is fixed. A total of 1090 cells are in
fact born in each worm, but exactly 131 of them die during
normal development. Mutations, however, can disturb this
process of programmed cell death. Some alter the mechanics
of death; others alter the marking of cells that are lined up to
die. The characterization of the genes involved and their comparison with homologous genes in other organisms have

shown that the mechanisms of programmed cell death are similar in many species. The characterization of these genes and of the signaling cascades that induce or prevent the cell death program is currently a very active area of research.[60] The decision "to be or not to be" is complex, the outcome of the integration of many signals. Once begun, the program involves specific cascades of proteases, enzymes that degrade other proteins and cellular structures.

The cells of a multicellular organism thus contain the seeds of their own destruction. When does the organism use these mechanisms? For the nematode, the answer is simple: programmed cell death is a key part of development. Together with cell division, differentiation, and migration, cell death contributes to morphogenesis.[61] Programmed cell death plays an essential role in metamorphosis in insects, but it also intervenes in the development of organisms in which the expression of the developmental program is less rigid and more regulated, such as in mammals. Here also, cell death can play a direct morphogenic role.

For example, humans have hands as we know them because of apoptosis: cell death takes place in the cells between the fingers. Otherwise, we would have webbed hands and feet. Programmed cell death can also have a more subtle role, such as during the development of the nervous system.[62] For example, in some species of birds, apoptosis leads to the female's losing those neurons that, in the male, control mating song production.[63] Death also comes to isolated neurons, owing to the absence of factors that enable neurons to survive. The function of this "death by default" is apparently to control the number of cells, by eliminating misplaced cells or favoring those that, having made the greatest number of contacts with other cells, will be most able to fulfil their functions.[64]

Cell death, which is the price that has to be paid for selection at the cellular level, also intervenes in the formation of the immune system. Programmed cell death is responsible for the disappearance of those lymphocytes that have learned by chance to react against the constituents of the organism itself instead of recognizing foreign bodies. Apoptosis also makes it possible to adapt the size and structure of organs to their function. For example, during the menstrual cycle, the number of cells that form the wall of the uterus changes. This is controlled partly by the number of cell divisions but also by the death of numerous cells. Programmed cell death is also used by the organism to destroy cells with altered genomes. As seen earlier, by escaping from this control and inhibiting the machinery of cell death, some cells become cancerous. Finally, starting the cell death program can also be the final stage in a pathological process—for example, in neurodegenerative diseases.[65]

The existence of "death genes" is paradoxical only from the point of view of the individual cell. True, no cell has any interest in possessing genes the sole function of which is to kill it. But for the organism as a whole, the existence of a program of cell death is highly advantageous. It can be used for very different objectives: producing a given form, adjusting the size of organs according to the needs of the organism, developing an effective immune system, or eliminating cells that have a poorly duplicated genome.

Programmed death is just one of the tools that organisms use to build themselves and carry out their functions. What programmed cell death does in one species will be the result of asymmetric cell division or cell migration in others. The logic of life lies not in the nature of genes but in the use organisms make of them to fulfil their functions. What counts is the end, and not the means used to achieve this end.

The mechanisms of cell death were invented very early on in the history of life. In bacteria, they enable populations of cells to conserve their genetic information: cells that have lost some of their information will either undergo apoptosis or will be killed by other bacteria.[66] However, we do not know whether programmed cell death in higher organisms is derived from these bacterial death programs, which vary substantially from species to species. Life has almost certainly invented programmed cell death several times over.

CHAPTER EIGHT

GENES AFFECTING BEHAVIOR

One of the most striking recent examples of the role genes play in complex behaviors is provided by research on the genetic bases of biological rhythms.[1] The study of rhythm genes is interesting because it poses point-blank the fundamental question of how a macroscopic function can be reduced to microscopic components: in this case, how can the complex mechanisms involved in the control of biological rhythms be reduced to the determinant action of one or a few genes? In what way do the products of these genes have a rhythmic structure?

We will focus on circadian rhythms, that is, biological functions that fluctuate according to the day/night cycle. These have been the most intensively studied of all biological rhythms and are the best understood. The path that led to the discovery of their genetic bases was long and complex. The first gene involved in the control of circadian rhythms was discovered nearly 30 years ago; it was the fruit fly gene named *period,* which codes for a protein called PER.[2] Some mutations of this gene lead to a complete loss of biological rhythms in the fly. On its own, this would not be sufficient evidence to conclude that *period* plays an essential role in rhythm control. As we have seen, a mutated gene can have effects that have no relation to its normal function. In the case of period, however, a number of additional arguments suggested that it does indeed play a key role in the control of circadian rhythms. Other mutations in the same gene do not abolish the rhythm but make it shorter or longer by a few hours. The amount of PER protein controls the rhythm: a sudden change in the expression of period can reset the fly's clock—the insect can be put at will into a particular phase of its circadian rhythm.[3] Both the amount of RNA transcribed from the *period* gene and the amount of PER protein show a cyclical variation, with a slight lag between the RNA peak and that of the protein.

Initial studies suggested that the PER protein negatively controls the transcription of its own gene: when the quantity of PER reaches a certain level, the protein migrates toward the nucleus,[4] thus leading to an inhibition of the transcription of *period.* This reduction in transcription leads to a reduction in the quantity of messenger RNA and in the rate of synthesis of PER, and thus in the level of the protein itself. If the level of PER is sufficiently low, the protein can no longer migrate to the nucleus—the transcription of period therefore increases, leading to a new accumulation of PER, its migration to the nucleus, and a new cycle.

According to this early scenario, the PER protein thus seems to form a fly "rhythm box": it not only controls its own expression but also, directly or indirectly, that of many other genes which show the same cyclical transcription rate. These other genes are responsible for the day/night variations in activity and function that can be observed in the fly.

Once researchers develop a relatively simple model, further experiments generally come along and make matters more complicated. A second gene, called *timeless,* has been discovered in the fly,[5] which when mutated also stops the fly's biological clock. The amount of the protein produced by *timeless* also oscillates with the day/night cycle and, like PER, when the TIM protein is at its highest level it is found concentrated in the cell nucleus. The simplest model that can explain these results and those previously reported for *period* involves a complex of both proteins, PER and TIM, which together form the fly's pacemaker.

When scientists tried to extend these observations to another insect, the Chinese oak silkmoth *Antheraea pernyi,* they encountered yet a new complication. The moth does indeed possess a *period* gene, the product of which—PER—also shows cyclic variation. Unfortunately, the moth PER remains in the cytoplasm throughout the cycle; thus, levels of PER cannot be the direct result of PER acting on its own transcription.[6] In this kind of situation, where an experiment tends to question previous results, scientists are rarely short of hypotheses that can explain away the problematic data. In the moth the PER protein may act, directly or indirectly, on another protein which would then have an effect on the transcription of *period.*

For a long time, the control of rhythms in mammals remained obscure. We know that in these organisms, a group of cells in the hypothalamus plays an essential role in the control of circadian rhythms.[7] Several research groups had suggested that a member of the family of proteins that respond to cyclic AMP—the CREB proteins—play a key role in controlling circadian rhythms.[8]

The CLOCK protein, the gene which has recently been sequenced in the mouse, was a better candidate for a key role in determining the circadian rhythm in this species, together with its partner, the BMAL1/MOP3 protein.[9] Part of the CLOCK protein (what molecular biologists call a domain) has a structure that is analogous to part of the PER protein. Does the *clock* gene play the same role in the mouse as *period* in the fruit fly? Yet again, experimental results did not confirm this apparently straightforward hypothesis: genes that are homologous to *period* have been subsequently isolated in the mouse and a gene homologous to *clock* has been discovered in *Drosophila*.[10]

On the basis of these and additional observations, general molecular models of circadian oscillators have recently been put forward which involve two interlocked feedback loops, each containing several molecular components.[11]

One of the key problems of the relation between genes and complex functions is how protein structure—the structure of gene products—embodies certain aspects of the complex function they perform; that is, to what extent is the macroscopic function already contained in the actual structure of the proteins coded by the genes that are required for the function to be expressed? What are the respective parts of "chance and necessity" in the design of molecular clocks? Rhythm genes provide part of the answer. They do not code for proteins that, in and of themselves, are rhythmic. Nevertheless, these are not just any old proteins. They share particular characteristics, such as a PAS domain, which is present in both the CLOCK-activating protein and in the PER-inhibiting protein, as well as in the proteins that control circadian rhythms in the mold *Neurospora*. This domain makes it possible for these "rhythm proteins" to interact and feedback indirectly onto their own synthesis. They thus generate simple sources of rhythm that the organism can subsequently perfect and integrate into complex molecular and cellular regulatory circuits.

Once such a cyclic self-regulatory system appeared, evolution conserved this kind of metronome. Other systems of self-regulation were doubtless possible: the dual processes of chance and evolution have meant that these systems either do not exist or are found only in other distant types of organisms.

For example, cyanobacteria (which, like plants, can directly metabolize light energy) use the same principle of a clock with several components, some playing a positive role, others tending to inhibit, but the molecular parts of the clock are very different from those present in animals.[12] When conservation of protein function has occurred—as in the case of insects and vertebrates—this does not imply that no changes have taken place. In fact, evolution has continually tinkered with the components and regulation of the various clocks, changing the number of parts and altering the modes of interaction and the quantity-control mechanisms.

One troubling result remains, however. In the fruit fly, *period* also controls a minute and very rapid rhythm—a very small cyclic variation in the frequency with which the male fly vibrates his wings during sexual behavior.[13] The explanation of the rhythmic role of *period* outlined above—which holds for the long rhythms of the day/night cycle and is based on a regulatory protein-synthesis feedback loop—cannot account for the control of rapid rhythms with a cycle of less than one minute. This might mean that *period* and PER do in fact have something special that makes them appropriate for controlling all sorts of rhythms. In this case, the model outlined above, in which genes only indirectly participate in setting up higher functions, would give way to the specter of the gene itself, which by its nature and that of its product would incarnate a complex function.

However, we should not worry too much: Jeffrey Hall and his co-workers have recently proposed a reassuring explanation for these enigmas. They suggest that *period* acts indirectly on the variation in the frequency of male wing vibration, as a result of its role in the development of the nervous control of wing movement.[14] The mutations that change the properties of PER might alter the development of this part of the nervous system and, indirectly, the rhythm with which the frequency of male wing vibration varies. This would imply that PER is not rhythm-become-protein but simply a protein that has been recruited twice by evolution to do the same job—producing biological rhythms—in two very different ways, one directly and the other indirectly.

GENES, SEXUALITY, AND PERSONALITY

Mutations in many genes can affect a wide variety of behaviors—if *fosB* is absent, maternal behavior disappears in mice, whereas *dunce* and *rutabaga* alter cyclic AMP metabolism and learning in the fly. However, the results of these experiments have to be considered very cautiously. As we saw, the characteristics of the knocked-out animals can depend upon the genetic background. Furthermore, it has recently been found that the same knockout strain, subjected to identical tests in different laboratories, can produce very different results.[15] This presumably indicates that undetected environmental differences—as was recently shown for food supply[16]—may have large behavioral consequences.

To find out whether genes are directly involved in the control of behavior, the study of isolated mutations in animals where genetic analysis is not yet highly developed is inappropriate because, in most cases, it is difficult if not impossible to straighten out the causal chain that links the mutation to the behavioral modification. The best approach—as in the previous case of genes controlling rhythm—is to choose an animal that has been intensely studied by geneticists, to describe a given behavior, and then to look for *all* the mutations that affect that behavior. Of course, to be of any fundamental interest, the behavior should be sufficiently complex for the findings to be extrapolated to other behaviors and other species. A genetic study of the bee's waggle-dance would be fascinating, but unfortunately our knowledge of bee genetics remains extremely limited.

A good example of this approach is the study of sexual behavior in the fruit fly.[17] When a male fly is placed in the presence of a female, he is attracted by her movement. He moves close to her, sometimes touching her with his legs, and comes into contact with the chemical attractants (pheromones) that she produces. The male begins to vibrate his wings, producing a buzzing love song with a species-specific frequency some components of which are indirectly controlled by the *period* gene. The amount of wing vibration is determined by the pheromones on the female. The female often moves away from the male, and he follows her, vibrating his wings. Other pheromones produced by the female then induce the male to lick her genitalia and then copulate. Although it appears that the female is passive, in fact her chemical signals largely determine the pattern and outcome of the male's behavior.[18] If the female has already copulated, she produces chemical and behavioral signals that discourage the male from continuing courtship. The fly's sexual behavior is thus highly complex, involving the exchange of visual, auditory, chemical, and tactile signals between the two partners.

This complexity explains the fact that many mutations have been found that alter the sexual behavior of fruit flies. The number of such mutations will doubtless grow—we are far from having described all the genes involved. Many mutations affect the fly's sexual behavior, although their primary target is another aspect of the insect's biology. Some of these effects are trivial—for example, mutations that reduce locomotor behavior: a male that is not very active will find it difficult to court. Similarly, but more interestingly, mutations that primarily affect vision, olfaction, or the emission or detection of auditory signals may all affect sexual behavior. Other mutations alter the frequency of male wing vibration, which makes the female less receptive to his advances. As well as these mutations, which affect key aspects of normal sexual behavior, there are also mutations that alter sexual dimorphism. Normally, there are a number of subtle differences between male and female fruit flies that go beyond genitalia. For example, the male fly has a specific muscle—called the muscle of Lawrence—which enables him to bend his abdomen. A mutation that affects the formation of this muscle in males might prevent copulation from taking place.

There are also differences in the brain structures that control the different phases of sexual behavior. In *Drosophila,* the brain is divided into different ganglia. Using complex and elegant genetic techniques, scientists can feminize or masculinize a part of the fly's body at will, including these ganglia. By feminizing certain parts of a male fly's brain, a fly can be produced that will court both males and females.[19] Although the exact nature of this bisexuality is unclear (the feminized male may be responding to male signals in an inappropriate way, or he may be simply unable to distinguish between males and females), this experiment shows that the brains of male and female flies have different functional characteristics and, no doubt, slightly different structures.

Males carrying the *fruitless* mutation also show high levels of male-male courtship behavior.[20] Furthermore, these males cannot perform most of courtship (singing, licking, and copulation) and show deformations in the muscle of Lawrence, although the relation between the two is unclear. It turns out that *fruitless* codes for at least seven different proteins, three of which are specific to males, three specific to females, and one of which is produced in both sexes. The *fruitless* gene appears to be involved in the cascade of genes that determine sex in the fruit fly, playing an important role in controlling the formation of the sexually dimorphic regions of the antennal lobe of the brain—a region that feminization studies suggest is involved in the detection of pheromones.[21]

Given the massive amount of work involved, the results from studies of the genes implicated in the fly's courtship behavior are somewhat disappointing. Many of the genes merely participate in the expression of a common, general factor—locomotor activity or sense organs. No gene can be considered as *the* gene responsible for any given phase of courtship. In fact, it is simply absurd to imagine that genes are responsible for behaviors—such a view would display a complete ignorance of the organization and functioning of organisms. On the other hand, the fact that the control of sexual behavior involves a specific nervous organization of the brain is very interesting.

The *fruitless* gene codes for a transcription factor that directly activates the expression of genes involved in the sexual differentiation of some nerve cells. This does not mean, however, that the causal chain which links gene and behavior is direct and immediate. This gene does not control male sexual behavior: at most, *fruitless* controls the formation of some of the neural structures that are involved in sexual behavior. The appearance of the gene did not determine the evolution of fruit fly sexual behavior; rather, it evolved at the same time, and as a transcription factor it was an effective coordinator of genetic activity—along with many other genes. The fruit fly's sexual behavior cannot be deduced from the sequence of *fruitless.* It is no more contained in this gene than traffic movement is contained in the structure of traffic lights. The product of this gene is merely one of the molecular components in a long chain of events which, starting with the complex ballet of macromolecules, transforms itself and crosses the different levels of organization of life—cell, tissue—ending up, at the macroscopic level, in the form of the behavior of the whole organism, in this case the fly. It would be absurd to reduce the sexual behavior of the fly to the properties of *fruitless.* The action of this gene—like that of all others—can be understood only after we have described all the different hierarchical levels of the organization of life through which its action passes, transforming itself at each step.

The communication between the different levels of organization takes place in both directions. In most animal species, such as humans or fruit flies, the chromosomal complement (and thus the nature of the genes) determines the sex of the individual. In other species, however, such as some reptiles and fish, environmental factors can cause individuals to change sex. This does not mean that genes play no role in sex determination in these species—it simply shows that information can travel from the environment to the organism and regulate gene expression. Genes do not determine the properties of the organism; they contribute to them. They are not the origin of organic characteristics, but they are involved in their realization.

These results from the humble fruit fly caused a major media storm because they apparently showed that the choice of sexual partner—sexual orientation—could be under genetic control. Furthermore, they appeared to support the findings from studies on the genetic bases of sexuality in a species that is much more interesting to all of us: human beings. In 1991

Simon LeVay, an openly gay neuroscientist, reported that the size of part of the hypothalamus varied with sexual orientation in homosexual and heterosexual men.[22] LeVay and much of the press suggested that the size of the hypothalamus determined homosexuality—the "gay brain." This view was supported by many in the gay community because it accorded with their personal experience of their sexuality—that they were "born this way"; sexuality was not a choice. However, the results are in fact far from clear-cut and have still not been replicated. The 19 gay men LeVay studied had all died of AIDS; his findings may have resulted from their pathology. And even if this were not the case, the change in this structure may have been the direct or indirect result of homosexual activity rather than its cause, or it may have resulted from some common factor related to both brain structure and sexuality. The arrow of causality can point both ways, or neither.

With a "gay brain," the "gay gene" could hardly be far behind. And indeed in 1993 a study found evidence for genetic control of sexual orientation in men.[23] The first part of the investigation was a classic piece of medical genetics: 40 family trees containing gay men were studied, and a correlation between homosexual orientation and the inheritance of markers on the X chromosome was found. Although the same team has confirmed this finding and has suggested that the same region is not linked with female homosexuality,[24] the validity of their results has been challenged on a number of grounds, and another team that attempted to replicate their finding has not been successful.[25]

Two conclusions flow from these studies. First, before any meaningful genetic analysis can be undertaken, the phenomenon under study—the phenotype—has to be clearly defined. Although homosexuality might seem like a fairly simple thing to describe, in fact it is very complicated. Being a homosexual today in San Francisco clearly involves different motivations and expresses itself in different ways than it does in Tehran. The same would be true if we compared any modern society with ancient Greece.

Second, many people found this study frightening because it suggested that sexuality might be at least partly genetically controlled. This fear had rational and irrational aspects to it. The rational component was that if homosexuality could be detected in one's genes, then reactionary ideologies could make use of this information to the detriment of the people involved.

The irrational aspect of this fear relates to the particular emotional charge given to genes and not to other biological determinants of behavior. For example, hormones can affect human behavior. Adrenaline accompanies aggression or fear and produces a set of physiological and psychological effects that we have all experienced. This effect is genetically determined in that genes control the synthesis of adrenaline and the molecular and cellular structures that process the hormonal signal. Nobody finds that fact disturbing. Some psychological differences between men and women are no doubt linked to hormonal effects (in particular during pregnancy), but no one finds that notion shocking either. These examples, like the others referred to above, show how the gene has become something onto which our fears and fantasies are projected. We will return to this topic later.

In fact, in all these cases (including the fruit fly) genetic "control" of behavior is highly indirect. The more complex the organism, the more developed the nervous system and the greater the role of learning. As we have seen in the case of memory, the nervous system's very structure—the precise nature of the synapses that connect the neurons—results from the activity of the neurons and therefore from the stimuli coming from the environment. In humans, behaviors are gradually acquired through contact with other humans; through language and thought, these behaviors take on symbolic value. Their components have meaning only in relation to the symbolic value that is attributed to them by the surrounding culture—a social phenomenon par excellence. In humans, although genes can be involved in making behaviors possible, they do not control them.

Over the last few years there has been a dramatic increase in the number of studies of the genetic control of human behavior, and more precisely of the genetic control of personality. Although human behaviors are above all culturally determined, genes might give them a certain coloration. Shaking hands—or rubbing noses—is not written in our genes; it is a practice acquired in childhood. Nevertheless, it is possible that a firm or a weak handshake could depend on the state of our genes, representing the role of genes in this behavior.

Before seeing whether various components of personality such as aggressiveness and anxiety are genetically determined, we first have to accept that personality traits can be described by a series of dimensions such as anxious/calm, responsible/impulsive, and so on. On this basis, several kinds of studies are possible, starting with measuring the personalities of identical and nonidentical twins. If a given personality trait has a genetic component, it will more often be present in pairs of identical twins than in pairs of nonidentical twins. On the other hand, if the trait is environmentally controlled, there will be no differences between identical and nonidentical twins raised in the same milieu. Results from such studies can then be filled out by studying identical and nonidentical twins who have been raised in *different* environments.

Studies along these lines appear to suggest that the genetic control of character traits is stronger than its environmental counterpart.[26] That is, whether identical twins are raised together or apart, they show significantly more personality similarities than do nonidentical twins who are raised together. Of course, the value of such observations depends entirely upon the validity of the traits chosen to define personality and the validity and reliability of the tests used to measure them. As a nonspecialist, I cannot judge these studies from these two decisive points of view. Whatever the case, the key weakness of this approach is that it does not help us to characterize the genes involved. This would only be possible by studying the transmission of such genes over several generations, rather than their distribution in a given generation. Although such studies may suggest that personality is under genetic control, they do not give us access to the genes involved.

The second approach is diametrically opposed to that of twin studies. As with research on genes associated with longevity, the starting point is a known gene that is suspected to play a role in the control of personality. This gene could be, for example, one that codes for a neurotransmitter receptor that pharmacological studies have shown has an influence on personality. The gene could also have been chosen because, in animals (generally in mice) mutation or inactivation leads to particular behavioral effects. The task is thus to see whether the gene exists in different forms and, if one of these forms is associated with a particular personality trait, with a behavioral difference.

This approach has not led to much progress in work on so-called genes for aggression.[27] On the other hand, success has recently been proclaimed in at least two cases. First, two research groups have shown that carrying one of two forms of the dopamine D4 receptor (dopamine is a central nervous system neurotransmitter) is associated with favorable or unfavorable attitudes toward novelty.[28] These studies were based on a series of investigations which showed that dopamine is implicated in the control of personality. Stimulating the release of dopamine leads to euphoria in humans, and in animals to an increase in exploratory behavior. In humans, carrying out a task that involves confronting novelty leads to an increase in activity in dopamine-rich regions of the brain. In animals, inactivating the genes that permit dopamine synthesis leads to apathy, as does inactivation of the D4 receptor gene.[29] The two forms of gene coding for the D4 receptor apparently give rise to proteins that fix dopamine differently: nerve activity in response to a given dose of dopamine will differ in individuals having one or the other type of D4 receptor.

The second result is the demonstration of a relation between anxiety and different forms of the gene that codes for the molecular transporter of another neurotransmitter, serotonin.[30] The two forms of the gene produce different quantities of the molecular transporter. As in the case of D4, a number of studies have shown a relation between serotonin and anxious behavior. Mice with mutations in the genes coding for certain forms of the serotonin receptor show an increase in aggression, which could be the consequence of increased anxiety. The brain regions activated in anxiety-producing situations are rich in serotonin and in its molecular transporter.

However, in both these cases—novelty-seeking and anxious behavior—the genetic differences thus revealed only explain part of the behavioral variability observed in the human population, which implies that other genes and other parameters (in particular environmental factors) play a role in determining these characters. Furthermore, in the case of the dopamine D4 receptor, the same genetic difference also seems to influence attention and tendencies to depression or drug dependency.[31] The fact that so many effects can be observed tends to cast doubt on the specificity of the observed phenotype.

The third, and most powerful, approach is that of isolating the genes involved in controlling personality and behavior without any preconceived ideas as to what exactly those genes might be. In other words, this approach extends to humans the methods that have been previously described with regard to the fruit fly and other animal species. This approach is difficult because, in humans, we can only follow genetic crosses between individuals—we cannot control them. Nevertheless, in the case of genes responsible for various diseases, this method has produced some success.

In the case of behavior, a supplementary difficulty rears its head: the traits under study are controlled not by one gene but by many, which renders the genetic analysis substantially more complex.[32] However, this difficulty can be overcome,[33] as demonstrated in the case of diabetes—a disease in which several susceptibility genes are involved.[34] Precise maps of the human genome have recently been established and the development of appropriate statistical methods has made it possible to pinpoint several such genes. Applied to mice, these methods have already shown the existence of genes involved in emotivity and alcohol- and drug-dependency.[35] Without waiting for similar results to be reported in humans but starting from the principle of a high degree of conservation of genes between species, the researchers who carried out these studies have already extended the validity of these observations to our species.

Finally, without being able to identify the gene involved, researchers can take advantage of the particular characteristics of genetic transmission and its anomalies to find support for the genetic control of behavior. Thus, a study has linked social skills with gender.[36] This work was carried out on young women with Turner syndrome—who only have one X chromosome instead of two. Such women differ from women with two X chromosomes in having a few anomalies in what are called the secondary sexual characters. However, they differ from one another in behavior, depending on whether their single X chromosome came from their father or their mother. Women with a maternal X chromosome and no paternal X find it more difficult to establish social relations than those with a paternal X chromosome and no maternal X. How can this be explained, if an X chromosome is an X chromosome, irrespective of whether it comes from the mother or the father?

The authors of the study put forward an explanation based on the concept of genomic imprinting. According to this hypothesis, some genes are affected by their passage in the male or female germ line and differ in their activity depending on whether they have been in a male or a female. The authors argue that there is a gene on the X chromosome that is necessary for normal social relations. This gene is inactivated when it passes via the female germ line, that is, when the X chromosome that carries the gene comes from the mother.

The implications of this study reach far wider than the behavior of women with Turner syndrome. It suggests that in the population as a whole, this gene will be active only in females, since males always receive their one X chromosome from their mother. This supposedly would explain why women have more highly developed social skills than men and why serious deficits in social interaction, such as autism, are found more frequently in males than in females.

This work is still at a preliminary stage. Nothing has been suggested about the nature of the gene involved. Indeed, there is no evidence that a single gene is involved—the study in fact implicates a portion of the X chromosome that carries several hundred genes. Furthermore, although the authors explain their results by genomic imprinting, this well-known phenomenon has never been described for the human X chromosome. Finally, there should be no hasty conclusions about the "sexism" of genes. If genes are involved, we can be sure of one thing: like the X chromosome which is present in both sexes, they are neither masculine nor feminine.

Geneticists have always tried to reduce the complexity of human behavior to the action of a handful of genes. Some earlier geneticists reported finding a gene for feeble-mindedness and, on the basis of this gene, explained all cases of mental subnormality in our species.[37] Today's geneticists are more prudent. But they do not always resist the temptation to oversimplify, in particular with regard to the interpretation of results rather than their description.

All such studies should be considered as preliminary—the reported results need to be confirmed. Whatever the outcome of subsequent investigations, it seems probable that different forms of genes will be found to be associated, in one way or another, with behavioral differences. But such a discovery will not mean that these genes are the only ones responsible for such behaviors. For instance, anxious behavior involves the activation of a series of specialized brain centers, thousands of neurons and regulatory pathways containing hundreds of different molecules. Serotonin and its transporter are just one part of these pathways.

If anxious behavior involves so many proteins and thus genes, why should interindividual variation apparently be based on only a few genes? The answer is not simple, but it would clearly be false to conclude that these molecular components are the most important part of the chain of events that lead to behavior. Perhaps these variable genes are those that code for the components that are found only in these pathways. The other genes involved would then participate in so many other biological processes that even a slight modification would have consequences too serious for the organism to tolerate. Or perhaps the consequences of such modifications, without being grave, would vary too much for natural selection to operate. Perhaps the variations we can observe are the only ones that are possible in the tight regulatory network that limits gene action. For reasons that are linked to the history of life and to the nature of the processes involved, the power of some genes has been less controlled and less restricted than that of the genes coding for the other components of these regulatory networks.

Is it really so shocking that some aspects of our behavior and our personality might depend on the form of a few of our genes? Is it not just as shocking that our mood can be altered by a handful of pills that the doctor has prescribed—often for a problem that has nothing to do with a personality disorder—through a series of side effects of medicines that in other respects are highly effective? Or by consuming a drug, be it legal or illegal? We have all experienced this kind of simple demonstration of the biological bases of mood and behavior. Is genetic control in some way less acceptable than pharmacological control? Accepting our genes is accepting the biological aspect of our human nature, nothing more.

GENES AND INTELLIGENCE

A great deal of research (and many books) have been devoted to the nature and origins of intelligence, yet we know very little about the molecular nature of the gene products—if any— that might be involved. The study of the genetic bases of this character in humans provides an excellent opportunity to see just how far it is scientifically legitimate to try to discover the molecular nature of a gene product involved in such complex processes.

The search for genes involved in human intelligence has in the past (both distant and recent) been marked by poor results that were sometimes downright fraudulent.[38] The racist or sexist motivations that often lay at the root of such work have largely contributed to discrediting this kind of study. It is easy enough to show the absurdity of some past studies and the many errors of interpretation that pepper the analysis of their results.[39]

Geneticists no longer try to demonstrate the existence of intelligence genes but rather study "the genetics of cognitive abilities."[40] This name change does not eradicate all the problems, including the fundamental difficulty of defining the object being studied. Scientists are far from unanimous that a general cognitive ability which can be directly measured by IQ tests even exists. Given that many studies have shown that general IQ levels have increased with the passing years, this would suggest that the cognitive abilities of the human population have improved rapidly.[41] IQ tests were invented by

Alfred Binet in order to orient children at school, and they only measure some aspects of what could be called "intelligence"—those that are approved by the education system.

Even if we accept that IQ is a good measure of at least something that is linked to cognitive abilities, the difficulties remain. A correlation between the IQs of parents and their children, or between the IQs of identical twins, even in old age, has been demonstrated.[42] But this does not help much: what causes the correlation—genes or environment? Detecting whether the correlation is stronger when there are more genes in common (as between parents and their biological, as opposed to adopted, children, or between identical, as opposed to nonidentical, twins) is not so easy.

Two key mistakes often mark such studies. The first is to suggest that if IQ is found to be partly inherited, then the genes involved directly control IQ. The link could in fact be very indirect. For example, suppose that some genes make children more interesting to their parents. The parents would then spend more time with their children and would pay more attention to their education, which would, in return, tend to favor development and raise their IQ score. In addition, the environment is not something totally exterior but is created—at least in part—by human beings themselves.[43] A higher IQ might be linked with genes that allow people to organize and improve their own environment, and thus increase their cognitive abilities.[44]

The second mistake is to conclude that, because IQ differences exist between two groups of humans and because within each group a genetic basis for IQ has been found, therefore genetic differences must be responsible for IQ differences between groups. In fact, environmental differences between the two groups can easily explain the observed differences even if,

within each group (where the environment is more homoge-
neous), environmental effects are weaker than genetic
effects.[45]

The debate on the heritability of IQ is still very much alive:
recent results suggest that the role of genes in determining IQ
scores is less than previous studies had found but that, on the
other hand, the effect of the prenatal or even later environment can be more important than previously thought.[46]

Robert Plomin's group has used another approach to identify genes that control intelligence. They have tried to find
whether high IQ is associated with particular forms (alleles) of
genes. Using 37 genetic markers on chromosome 6, the researchers found that very bright children tend to have a particular allele coding for an insulin-like growth factor receptor.[47] The role of growth factors was discussed earlier, both in
the context of brain development and, more generally, the activation of cell division. If Plomin's result were to be
confirmed, what would it tell us about the biology of intelligence? Simply that cell division in neuronal precursors is important for the developing brain. What a surprise!

Throughout this book, I have tried to steer away from the
classic genetic approach that correlates differences between
individuals with differences in alleles in order to focus on the
common action of the genes present in all the members of a
given species, and thus on the way that they enable the construction and functioning of life. Applying this method to the
question of the genetic bases of intelligence is particularly en-
lightening. Rather than focusing on variation in genes that
might explain possible individual differences in intelligence,
surely it is more useful to study the genes that have enabled
human intelligence to appear. In fact, there is no reason to
suppose that the genetic variations which, in conjunction with

the environment, might potentially explain individual differences are those same variations that have led to the evolution of human intelligence.[48] A simple example will underline this point, the necessary distinction between the factors involved in interindividual differences in one characteristic trait and those controlling the realization of this trait. Humans have five digits (fingers or toes) on each member (hand or foot). There are some—rare—genetic conditions that modify this number. However, variations in the number of digits are more often due to developmental anomalies, such as those provoked by the drug thalidomide. A geneticist studying the transmission of digit number from parent to child, in the same way as others study the transmission of IQ, would have to draw the inevitable conclusion: variation in digit number is caused by the environment, and only occasionally by genes. And yet, the fact that we have five fingers is clearly coded in our genes. But not in a simple, direct way that is easily alterable by a mutation in one or another of these genes. There is no gene for a finger, or for finger number. The number of fingers results from the indirect action of many genes that probably mutually compensate each other for slight variations in their activity.

Are there genes that are involved in the development of human cognitive abilities, and if so, what is their nature? Human intelligence clearly depends on the social environment, on the presence of other humans. The many descriptions of feral children who have been abandoned in early childhood and reared in the absence of any communication show the vital role of society. Nevertheless, these observations do not contradict the possibility that genes could also be important and necessary for the acquisition of specifically human cognitive abilities.

The factors that have enabled us to make tools, to speak,

and to begin the long march toward civilization are probably contained in our genes but in a "cryptic" form. Is it possible to go further and try to describe the nature of the genes involved? Unfortunately, no. At the moment, we know nothing about the genes that make us human.

In order to understand the problem better, rather than remaining on the shifting ground of intelligence, I will concentrate on the role of genes in language acquisition. The appearance and development of language was decisive: it accompanied and enabled the development of human intelligence and the rise of our species.

A number of debates still excite linguists. Is there a universal grammar, as Noam Chomsky proposed? Are there so-called cognitive modules that function autonomously in the brain? Different genetic explanations would have to be put forward depending on the answers to these questions. If the existence of cognitive modules were to be confirmed, one could imagine the existence of genes specialized in the construction of these modules.

There is one deeply troubling observation in favor of the modular organization of cognition and language and their genetic control. The linguist M. Gopnik has described the case of a family in which some children have difficulties learning their native language. The large size of this family made it possible to show that the deficit was probably linked to a single dominant mutation.[49] Curiously, the deficit was highly specific: affected individuals had difficulty learning only some grammatical rules—how to put nouns in the plural and how to conjugate verb tenses.

Although the attribution of the deficit to a single dominant gene mutation has recently been confirmed,[50] the model proposed by Gopnik and his co-workers has been criticized.[51]

The critics interpret the same observations as the sign of a more general defect in language production. This seems to me to be more probable, given what modern biology tells us about gene action. The idea that there are "language genes," as suggested by neurophysiologists, linguists, and psychologists, is both naive and reductionist. The same can be said for the idea that there are genes for reading and writing, even if a large number of studies suggest that reading difficulties may be associated with particular forms of genes.[52] Do the advocates of such ideas really think that they will find the explanation of a higher function like language in the structure of genes and in the proteins they code for? If so, they clearly do not realize that proteins—the immediate products of genes—do not act in isolation but participate in the formation of complex networks and structures that are integrated into the hierarchical organization of life.[53]

If we want to discover the role of genes in the evolution of language, the best approach will probably be to study the development and organization of those regions of the brain that constitute the language centers and of the complex neural structures which preceded and allowed the formation of language.[54] It would be absurd to try to study directly the genes for language. If, one day, we are able to characterize the genes that are necessary for language acquisition in humans, I would be very surprised if these genes did not code for components of intra- and intercellular signaling pathways found in the neurons of the brain regions that are specialized in the production and interpretation of auditory messages. The explanation of language can no more be sought in the structure of these components than the voice that comes out of a radio can be explained by the transistors in the radio. And yet, without these genes, without the characteristics that their products enable the nerve cells to acquire, language would never have evolved.

GENES AND ALTRUISM

The final example that we will consider of a "higher" function for which a genetic basis has been proposed is altruism,[55] or, more exactly, altruistic behavior. The existence of genes that effect altruistic behavior was first proposed in a field outside the context of classic genetic studies.

The genes for altruistic behavior are the result of research on social insects and owe their origin to the work of sociobiologists.[56] Darwin and his successors conceived of natural selection as a struggle for reproduction. The existence of altruistic behavior in animals—particular obvious in social insects where workers do not produce their own offspring but rear those of the queen—posed a major challenge for Darwinian theory. In termites, some individuals even play the role of kamikazes, literally blowing themselves up as they spray threatening predatory insects with a toxic liquid. How is it that such behaviors have been conserved by natural selection, when they are clearly harmful for the individuals that perform them and thereby diminish their reproductive success? Altruistic behaviors also exist in other animals apart from social insects. Some birds emit alarm cries when they notice a predator. These cries improve the survival chances of all the threatened animals, but they also enable the predator to localize the particular bird that produced the cries.

William D. Hamilton was the first to find an explanation that gave altruistic behaviors a selective value.[57] Building on the insight of G. C. Williams,[58] he redefined the objective of life: rather than produce offspring, the name of the fitness game is to transmit one's genes to the next generation. Not because genes are something sacred, something superior to the rest of the organism, but simply because they enable reproduction to take place. By helping out one's kin, one increases the chances that genes shared in common will remain in the gene pool of the next generation. This process is called, appropriately, kin selection. The fitness advantage of altruistic behaviors for the transmission of genes can be demonstrated quantitatively in social insects, where the rigidity of behavior and the simplicity of the rules of reproduction make such a study relatively simple.

Are there genetically determined altruistic behaviors in humans? Some researchers have suggested that morals and ethics are simply the extension of a genetically controlled altruism, the cultural expression of a natural behavior. It seems to me that such a hypothesis takes little account of the sea change introduced into human behavior by the acquisition of language and symbolic representation.

In addition, not all social behaviors that imply cooperation between the members of a given animal group are necessarily genetically controlled altruistic behaviors. They may simply be long-term strategies, an exchange of services that is mutually beneficial to all concerned.[59] It is even possible that some animal behaviors that have previously been considered altruistic—such as predator look-out behavior—may in fact be selfish and thus directly advantageous for those animals that perform them.[60]

But let us accept for a moment that altruistic behaviors really do exist and that some of them are genetically controlled. The existence of such behaviors poses two problems. The first involves understanding how an animal recognizes individuals that should be the object of its altruistic behaviors, that is, kin with which it shares genes. This ability turns out to be widespread in the animal kingdom.[61] The simplest explanation is that an animal recognizes its close relations because they resemble, according to one criterion or another, the first organisms with which it was in contact (father, mother, brothers, sisters). In fact, it is merely necessary for closely related animals to live together or close together for any altruistic behavior—even a nondirected one—to benefit above all those individuals that carry the same gene form and thus to favor the spread of this particular form throughout the population.

The second problem—more difficult to resolve—involves understanding how a particular form of a gene can favor an altruistic behavior. What link could possibly exist between a gene that codes for a protein and the complex whole that is an altruistic act? It would probably be wiser not to try to make such a link. The existence of genetic variation that might favor altruistic behavior does not imply the existence of genes that directly control such behaviors.[62] An altruistic act is a complex biological process, involving several different regions of the brain, thousands of neurons, hundreds of types of ion channels, receptors, and protein kinases. Furthermore, altruistic behavior cannot be isolated from the totality of behaviors of which it forms a part—complex behaviors in which periods of aggression often alternate with periods of altruism.

If some altruistic behaviors can be shown to be genetically controlled, it would only mean that among the hundreds or thousands of genes involved, for reasons that we will no doubt have difficulty discovering, the variation of one particular gene leads to a slight imbalance in the behaviors, giving them a more marked altruistic aspect. Neither this gene nor any of the others is responsible for the behaviors. Even less so is it responsible for the altruistic aspect of the behaviors. It is simply a pivotal point at which a slight imbalance in gene action can have major consequences, the feather that makes the scales tip in favor of an altruistic behavior.

At the end of the 1930s, Conrad Waddington proposed the metaphor of a landscape to describe gene action during development.[63] This metaphor fell into disuse but has recently been adopted by some geneticists.[64] Genes contribute to the landscape in which the biological process takes place, creating mountains and valleys. The route followed by development or by personality formation is analogous to that followed by a river. It depends directly on the landscape, indirectly on gene action. Sometimes a dike only has to be raised by a few meters for a river to change its course and run down to the sea several thousand kilometers away from its natural mouth. This image has the advantage of showing that, in a complex structure, a small change in constraints can have massive consequences. Thus the modification of a single gene that, together with others, is involved in the expression of altruistic behaviors could have important effects. Nevertheless, it would be an exaggeration to say that this metaphor tells us much about gene action. The use of such metaphors is only an elegant way of hiding our ignorance.

CHAPTER NINE

WHITHER GENETIC DETERMINISM?

At the end of this journey through recent results from biology where researchers are trying to understand how genes intervene in processes as complex as the control of biological rhythms, development, aging, intelligence, or altruism, we now have enough data at our command to put forward a less naive view of gene function and reconsider the notion of genetic determinism—the existence of a deterministic link between genes and characters.

This determinism comes out of our quest weakened and threadbare. Not because, as some would have us believe, chaos and disorder exist at the molecular level—in fact, life is virtually always based on a strictly regimented, structural, and dynamic order—but because of the hierarchical structure of reality. As the physicist P. W. Anderson put it in 1972, "The ability to reduce everything to simple fundamental laws does not imply the ability to start from those laws and reconstruct the universe."[1] In the case of biology, determinism is undermined by the organization of the molecular components encoded by the genes that make up organisms.

GENES HAVE MULTIPLE EFFECTS

The first characteristic of gene action which stands in the way of a deterministic conception of it has been called genetic promiscuity:[2] a given gene can be involved in several different biological processes. An extreme example is the case of proteins that have a double function—for instance, enzymes that are involved in both metabolism and the regulation of transcription.[3] Such moonlighting proteins are less frequent than those with a unique activity that is used in different processes. Some genes intervene at several points in development to carry out similar functions but in different contexts; for example, the genes coding for the NF-AT factors intervene in both the immune response and in the development of the heart valves, and some memory-formation genes also control the storage of sugars in the organism.

As Denis Duboule has pointed out, this multifunctionality of genes means that we need a different approach to the problem of genetic determinism and in particular to the determinism of form.[4] The fact that we have five fingers on each hand is the result of the action of homeotic genes. It can be said that these genes determine the number of fingers we have. But why do we not have four or six fingers? What are the constraints that maintain the number at five? They may not in fact be linked to the construction of the hand but be imposed by other functions carried out by homeotic genes in constructing the organism, in the digestive tract or urogenital organs. From this point of view, the number of fingers would be the result of constraints that act at a different place. The consequence of this genic multifunctionality, of this tinkering with developmental genes, is that as the organization of life became more complex, evolution became more and more constrained. As a result, evolutionary transformations became less supple and gradual and increasingly "punctuated."[5]

From its very beginning, classical genetics showed the existence of pleiotropic gene action. Molecular biologists have merely explained the reason for this phenomenon and shown its extent. This multiple action of genes, their involvement in very different processes, is not arbitrary: it is because the genes and their products have particular properties that they carry out these multiple functions. For example, the Notch proteins and the various other protein components of this signaling pathway make it possible for neighboring cells to exchange information. Such exchanges make it possible to program differential but complementary development in different cells. Once these regulatory networks have been set up, they can be adapted to an infinite number of uses during development: one only has to reflect on the wide variety of organs in which several different cell types coexist in clearly defined proportions. In addition, because the Notch proteins are at the membrane, they have also been used to control other processes linked with cell-to-cell interactions, such as the growth of neurites (the extensions of neural cells).[6]

This gene "recycling" in order to carry out different functions during development explains the fact that, in humans, mutations in single genes can lead to very complex syndromes with defects in the brain, face, limbs, heart, kidneys, digestive tract, and so on.[7] The list of symptoms would doubtless be even longer if genetic redundancy—the second important limit to individual gene action—did not mask many defects.

Genes Act in Cohorts

The multiplicity of genes also makes their individual role all the more confused. We have seen that the genomes of organisms like the vertebrates originated from the repeated duplication of the genomes of simpler multicellular organisms. Each gene present in the fruit fly corresponds to a family of mammalian genes, each with similar functions. Although these functions are not identical within a gene family, they are sufficiently close for there to be a partial redundancy, and thus a possible compensation, between the functions carried out by each of these genes. In fact, functional compensation is caused not only by the existence of multigene families. Knockout experiments show it is a more widespread phenomenon. It can intervene within a cell, or between cells within an organ or an organism. As well as functional compensation, there is an adaptive compensation that, during development, makes up for genetic deficiencies. These observations, made using the most sophisticated techniques of molecular biology, are merely the rediscovery of the properties of plasticity and regenerativity of organic systems that were described long before biology existed as a science.

An image taken from physical chemistry—initially proposed by Waddington—can be used to describe this ability of organisms to cushion the impact of genetic variations.[8] One of the characteristics of the internal milieu of organisms is that it is buffered—for instance, its physicochemical properties remain stable despite the addition of large quantities of acids or alkalis. Similarly, the genome is a buffered structure in which variations in gene functioning are cushioned. The totality of genes has evolved such that the consequences of genetic variation are minimized.

Some examples of compensation show that the organism can even free itself from anatomical and genetic constraints. The case of blind people is particularly striking: they often develop a greater touch sensitivity than normally sighted people. It has recently been shown that the occipital brain regions, which usually process visual information, are used by blind people to process tactile stimuli—a good example of compensation that cannot be interpreted in terms of molecules or genes.[9] Genes did intervene in this process, through the construction of distinct brain regions that optimize the processing of visual and tactile information. The study of blind people shows that the organism is not the prisoner of its genes, or of the structures that have been formed under genetic control.

Finally, the complex signaling and regulatory pathways and the organization of these pathways in networks make it very difficult to determine the particular role of each gene. The major changes observed during evolution are more the consequence of the reorganization of such networks than the modification of the protein links that form them.[10] Sometimes, such changes may have been produced by a mere change in the level of expression of these links or the period during which these proteins are active. These characteristics of the functional organization of genes explain the fact that, despite the fundamental role they play in the construction and function of organisms, it is nevertheless very difficult to state precisely which genes are involved in a given biological process.

The only exceptions to this are a few cases where natural selection has acted with exceptional force. In these relatively rare cases where the action of a single gene and its product stand out from the functional and regulatory network, the biological process in which the gene is involved becomes fragile, because a single mutation can have such deleterious effects. Why are there such situations, where a gene plays such an important role? Is it simply that, in a network, there always remain weak links, points from which the whole fabric can unwind? Do these situations correspond to particular phases of evolution, just after a new process has developed and before the action of the gene has been buffered by the whole genome?

The counterexample of aging is particularly revealing: in this case, where natural selection has had little effect, we have seen how the whole physiological process depends on the properties of genes, without there being "genes for aging." An image, or comparison, may make things clearer. An organism in its environment is similar to a boat on the sea, blown and buffeted this way and that. But there are two kinds of boats. There are those that are used for transporting passengers or goods. If they are well built, then, given some sea-faring knowledge on the part of the sailors, they will resist the dangers of the sea. There will be a few incidents affecting different parts of the structure—the hull, the motor, and so on—without it being possible to say that any particular part is the weak link. Every bit of the boat makes it seaworthy, each bit playing an equal role. There would be little point in optimizing any one of these components in the hope of avoiding a rare event. But there are also racing yachts that have been specifically designed to increase performance by enhancing a few key components. These elements will make the yacht go very fast, but they also make it more fragile. The other components of the yacht, made resistant by experience, will not fail in their appointed task. Does this mean that these other components are less important for the boat to go fast than the newly modified one? The answer is yes if the yacht is compared to its competitors, no in absolute terms. As the years pass, and experience grows, the modified parts that were weak and fragile in the earliest yachts will become more solid and more reliable. Their structure and their function will have been optimized—and then designers will no longer concentrate on these factors. These components will no longer reveal themselves by failure—other components that have been changed to increase performance will now be in this position.

The genes that are revealed by genetic analysis may be those that have been made more fragile by natural selection's recent modification of their activity or regulation. We should not forget the existence of all the other genes which have a silent, consolidated action and which make it possible for the observed functions to be effected.

Gene Action and Organic Hierarchy

The action of gene products is expressed only indirectly through an organizational and structural organic hierarchy—protein machines, organelles, cells, tissues, organs, organisms, and populations.[11] This fact undermines genetic determinism more than the previous arguments, so much that the very word "determinism" is no longer appropriate to describe gene function. There is a precise causal chain linking the product of a gene to the actions of that gene within the organism. But this causal chain passes through different levels of organization. At each level the chain is transformed and obeys different rules. The elementary rules—those that apply at the molecular level—are insufficient to understand the macroscopic effects that can occur through the modification of a protein coded by a gene. Each level of life's organization imposes its own order, its own logic.

This is already true at the simplest level of organization, that of the protein machines that catalyze complex cellular reactions such as the replication of DNA or its copying into RNA. Where isolated enzymes or proteins merely take into account the concentration of reactive molecules present in the medium, these complexes channel such reactions in a precise direction, limiting the diffusion of the molecules between each elementary reaction. Without violating any principle of chemistry but simply by limiting the range of possible outcomes, these machines orient the elementary chemical operations of life, making them more efficient.[12]

This example was solely intended to show how hierarchically structured organization can orient and transform the action of elementary agents such as proteins and enzymes. Not all levels of organic organization have the same importance. Although that of the organism is, of course, essential, the cellular level is equally fundamental. The rediscovery that the cell represents a major level in the integration of biological processes (and thus in their explanation) was probably the most important change that has taken place in molecular biology since its rise in the 1950s. And, as with virtually all important conceptual changes, most people did not even notice.[13] In the 1960s, cell biology appeared doomed to disappear in the not-too-distant future. The dream of the molecular biologists was to explain directly the characteristics of organisms on the basis of the properties of the proteins that formed them. And yet, since the 1970s, cell biology has undergone an extraordinary renaissance. Not merely because of the development of new techniques but mainly because the cell turned out to be fundamental on two levels. First, far from being a bag of enzymes, it is a highly organized structure, within which molecules and information circulate. Second, biological phenomena as complex as the immune response, development, or the appearance of cancer can be explained only if the cell is seen to integrate all the information it gathers from itself and its environment and to enjoy a certain degree of autonomy.

In order to understand the logic of life, we need to understand its structural hierarchy, an order that develops out of its molecular components but which imposes higher rules upon these components. What possible reason could there be for programmed cell death without the existence of the organism, of its "will" to survive (for want of a better term)? How can we explain at any other level than that of the organism, by reasons other than the necessity for this organism to reproduce efficiently, the fact that, in related animal species, development follows different paths, using different molecular and cellular pathways to arrive at the same macroscopic result, while sacrificing some cells along the way if need be? There is nothing miraculous in these abilities of the organism; there is no hidden purpose or teleology at the heart of life. It is simply the case that the organisms we study today are those that have inherited genes that make their reproduction efficient.[14]

We can read back from the phenotype to the gene, following the causal chain that explains the character by the properties of the gene and its product. This causal chain is perfect, and each of its links can be precisely described. But if, on the other hand, we start from the gene and try to predict its effects, the diffraction of the causal chain at each increase in the level of organization makes it difficult to predict anything at all.

This vision of gene action as a determinism that is fractured by the hierarchical structure of life means that it is not necessary to choose between the rigid genetic determinism proposed by some biologists and the rejection of a precise role for genes in development or behavior that is advocated by others. Both are true, but at different levels of organization. It also explains the paradox raised by some physicists and mathematicians who have become interested in the study of life: the amount of information contained in the genes of any organism is insufficient to explain its organization and functioning. This would be true if organisms had only one level of organization—the molecular. But that is not the case: organisms are made up of a hierarchy of organizational levels. The relation between the complexity of a hierarchical system and that of its components is not linear—the former can grow quicker than the latter. In fact, apart from some pioneer studies such as those of Stuart Kauffman,[15] the term complexity is (too) often used to hide our ignorance of the functional rules that the existence of a hierarchical system imposes on its components.

This conception of the role of genes represents an enormous challenge to all those who want to model the function and evolution of organisms. At the beginning of this book, I explained how population geneticists use an out-of-date conception of the gene, one that was developed before the advent of molecular biology. This is not because population geneticists are lazy but simply because the molecular vision of the gene would require them to include the hierarchical organization of life in their models. They have chosen to systematically ignore this up until now.

The population geneticists are not alone, and most of those scientists who have tried to model the functioning of organisms did the same. Can this resistance be overcome? I am not competent to answer. But it will certainly not be easy: successful models, particularly in physics, have always dealt with homogenous chunks of reality. Modeling the interface between different levels of organization and the diffraction of causal chains as they pass through these interfaces will be a major challenge.

The characteristics of gene action make it impossible to group genes in functional categories—the genes of cancer, developmental genes—or to reduce the complex properties of organisms directly to the characteristics of one or a limited number of genes.[16] One way of going beyond our present limited understanding would be to use specifically-designed experimental approaches, such as the two-hybrid technique, to unravel the complex circuits and networks in which proteins act. Another approach, "functional post-genomics," aims to provide a complete description of the activity of the genes in a given cell or tissue.[17] Its advocates hope that new principles of biological organization will emerge by themselves from the mere observation of gene global functioning.[18] It is too early to know whether these hopes will be fulfilled. What is clear is that a strong motivation behind these different approaches is precisely the difficulties of interpreting individual gene action that we have emphasized.

FEAR OF GENES

A final advantage of this new conception of the role of genes is that it should reassure those who are alarmed by the power of genes. Our genes do not code for our behaviors or our thoughts, but without genes these behaviors and thoughts would not exist. The many examples outlined in this book show that human beings are no more prisoners of their genes than the painter is a prisoner of his or her palette or the architect of the laws of gravity.

So why are we so afraid of our genes? They are nothing more than components of our bodies. We accept that our actions, our behaviors, are conditioned by the state of our bodies. At the beginning of this book there is a quotation from Condorcet, which states that profound thoughts are at least as tiring, if not more so, than physical effort. This shows, somewhat crudely, that "intellectual organs" are necessary for this thinking to take place. Would it mean anything more to say that profound thoughts are possible because of the action of our genes?

Nevertheless, it does appear that our genes have a more negative image and are thought to be more constraining than our bodies. This can be shown by three examples, two of which have already been mentioned.

First, many drugs that act by imitating or opposing the action of natural neurotransmitters alter mood and behavior and thus affect personality. This shows that our mood, our personality, depends directly on the action of neurotransmitters. Imagine—as some experiments already suggest—that the form of some genes which intervene in the synthesis of these neurotransmitters or their mechanisms of action were found to have an effect on mood or personality. How would this restrict our freedom any more than the straightforward effect of neurotransmitters? In what way would such genetic factors make us biological slaves?

The second example relates directly to medicine. For decades, a clear link has been established between strokes and a gradual elevation of blood pressure as we age. Many of us will have to take drugs when we are older to lower our blood pressure and to reduce the risk of circulatory problems. Biotech laboratories are currently trying to find the genes that make us susceptible to high blood pressure—that is, those genes which, when they are present in a certain form, make it more likely that we will develop hypertension. It is very likely that such genes will indeed be characterized and that, in a few years, it will be possible to know who is predisposed to high blood pressure. Once this risk has been identified, we may choose to adopt an appropriate life style, eating habits, and medication, according to the state of our genes. What should worry us is not the possibility that one day, by understanding our genes better, we will be able to remedy our health problems earlier, in a less constraining and more efficient way, but rather the current situation. When high blood pressure is diagnosed today, it is too late for prevention—our arteries are already clogged up.

The third example is that described earlier in the chapter on the surprising results obtained by gene knockouts. The *fosB* gene is one of many genes that are involved in intracellular signaling. If the gene is absent, there is a very specific effect—the loss of maternal instinct. Many people have been struck by this finding: maternal instinct apparently depends on the action of a single gene!

What is truly shocking is the suggestion that there is a simple, direct relation between the *fosB* gene and maternal behavior. The effect of the *fosB* gene can equally be interpreted differently, or rather, we can avoid making a short-cut between the *fosB* gene and maternal instinct and instead unravel the long causal chain that links a mutation in the gene with the observed effects. The *fosB* gene codes for a protein that is an important component of the intracellular signaling pathways. The pathway that the FosB protein is involved in is necessary, among other things, for the organization and functioning of the neurons that form one of the centers of the hypothalamus, a structure situated at the bottom of the brain which controls feeding and emotional behaviors. This structure is important for maternal behavior: neurobiologists have shown that if it is destroyed experimentally, this behavior disappears. This model explains why the FosB protein is necessary for the expression of maternal behavior, and it is confirmed by the fact that mutations in other genes coding for proteins that are important for the development or functioning of the same hypothalamic structures have effects that are similar to those resulting from the deletion of *fosB*.[19] Once again, this example shows that the power of our genes is nothing more than the power of our bodies.

So, why does the power of genes make us so afraid? Why does it appear so constraining? Why does it have this air of fatality? Why are we afraid of our genes? The first and least important reason is that our genes not only constrain ourselves, they also constrain our offspring. A genetic diagnosis not only affects one person, it affects several. However, although this is true, it is mainly so for a small number of dominant mutations affecting a single gene—for example, Huntington's disease. Most genetic diagnoses that will be made in the future will involve testing for susceptibility genes to certain diseases. For each disease, there will be many such genes. The risk will be linked to a particular combination of genetic forms and events that took place during life. Such a combination will not be transmitted to the next generation.

It is obvious that the fear of genes goes far beyond the fear of transmitting diseases or defects to one's children. This fear is intimately linked to the history of biology and of genetics and to the metaphors biologists have used to explain gene action. If we are afraid of our genes, it is because they appear as the fundamental constituents of all living matter, the one thing around which everything else is organized.

Genes Are the Products of Life's History

Such an emphasis on the role of genes to the detriment of all other aspects of life no doubt flows from the reductionist tradition of modern science that has always tried to explain phenomena at the simplest level. This approach paid off in physics, starting with the Scientific Revolution, and ever since that time other disciplines have tried to copy these successes. When genes were discovered, many people agreed with Hermann Muller, who called them the biologists' "atoms."[20]

The discovery that genes contained the complete structure of proteins in the form of a code made it possible to consider the genome as a "text." How pertinent is this particular metaphor? In what way does it reflect the predominance of genes? The genome has two remarkable characteristics that distinguish it from all other components of life: its stability over the life of the organism—and even more so over the generations, despite the mutations that allow organisms to evolve—and its unity, the fact that it contains the information necessary for the synthesis of all the organism's proteins and thus of all its active constituent parts.

But on its own, the genome is simply a dead memory. The real biological agents are the proteins coded by the genome. By their specific action, and through the hierarchical organization that characterizes all life, proteins give organisms their distinguishing characteristics. The stability and "wholeness" of the genome is itself the result of the action of proteins (such as the helicase, which, when it malfunctions, leads to the accelerated aging observed in Werner syndrome) that continuously correct and repair the genome.

Despite these different but complementary roles of the two fundamental components of life, the balance still seems to tip in favor of genes: their structure is not determined by the action of proteins but comes from the faithful copying of the structure of other genes. On the other hand, genes determine the structure of proteins, and thus their function.

The control of protein structure by genes is exercised in two phases: the first, which has been perfectly described, involves a strict correspondence between the fine structure of the gene (a sequence of nucleotides) and the equally precise structure of the protein (a clearly defined series of amino acids). But the newly synthesized polypeptide chain is simply a molecule without form or function. The formation of the active protein is the result of the spontaneous folding of this long polypeptide chain, even if, as we have seen, the process may be helped along by molecular chaperones.

The genetic control of protein structure is thus very particular: protein folding takes place spontaneously, on the basis of kinetic and thermodynamic laws. The gene does not contain protein structure: history and natural selection have simply established a relation—virtually perfect, because protein folding is a very effective process—between a given gene structure and a particular protein structure. The genetic control of protein structure is not a logical determinism but a historical determinism, which has gradually developed and taken form over the history of life.

This link between genes and proteins is not primordial, it is not consubstantial with life. It followed the appearance of the first life forms. As we saw in the opening section, if the current scenarios of the origin of life are to be believed, the world of DNA and proteins that we know today was preceded by a living RNA world.[21] In this RNA world, the RNA molecule was the bearer of properties that today are separate and linked to two different macromolecules—self-replication is now the role of DNA, whereas proteins ensure precise functions in the organism. DNA and genes were among evolution's late inventions, a means for life forms to improve the reproduction of their active molecules.

Whether genes are studied at the most elementary level or whether we follow their appearance in the course of the history of life, they are merely a tool invented by life to effectively reproduce the structure of its active agents. Writing did not precede language and is not superior to it. However, the invention of writing substantially increased the power and effectiveness of language. Similarly, genes are not superior to RNA molecules or to proteins. However, genes and the genetic code have provided cells with a remarkably effective way of synthesizing their biological catalysts and have made it possible for life forms to acquire the richness and diversity that surrounds us today.

However, we can regret the passing of the RNA world in which two fundamental characteristics of life—self-replication and the organization of complex molecular structures—were expressed by a single type of macromolecule. If the development of thought and of science were possible in the RNA world (which is unlikely), biologists and philosophers would not have wasted their time in sterile debates as to the relative importance of genes and other components of life.

CHAPTER TEN

Human Evolution and Eugenics

Human genetics frightens people. And not without reason: at certain periods in our history, it has been used as part of a policy of racial discrimination.[1] People fear that human genetics will reveal fundamental differences between individuals or between human "races."

However, the results of studies of the distribution of genes in human populations show that such fears are unjustified. First of all, a widespread illusion must be discarded: human genetics tells us nothing about the genes that make us human—those that enabled the acquisition of cognitive and linguistic abilities and the rise of civilization. The nature of these genes remains a mystery, and it is thus impossible to study them. They are, no doubt, numerous and not particularly human. The development of our species is more the result of changes in gene expression than the acquisition of new, specifically human genetic information. The study of fossil skeletons and the behavior of our ancestors tells us more about our evolution than the study of our genes. Understanding the genes involved in the appearance of our species would not replace the need to understand the brain structures those genes created and that are responsible, in particular, for the acquisition and expression of language.

Human genetics tells us nothing, either, about any possible qualitative differences between human "races." Quite the opposite: it shows that, despite appearances, races do not exist in the human species in the same way they do in some animal

species. So-called interracial differences do not reflect the genetic variability between humans—scientists who have carried
out studies of the distribution of blood groups or other similar
genetic and biochemical markers have been surprised to discover that different human populations generally contain the
same alleles of a given gene.[2] Human groups are distinguished
by the frequency of different alleles, not their presence or absence. This shows that the mix of genes within the human
population is very strong, and that no human group is genetically isolated. Humanity is genetically whole.

How can this result be reconciled with the evidence that,
following their geographic origin, human beings can be so different for physical characters such as skin color, hair distribution, or eye form? Despite all their training, are geneticists
less competent than your average immigration officer in distinguishing among ethnic groups? In fact, the two types
of genes are not the same and are not subject to the same
selection pressures. Climate and environment exert strong selection pressures on the handful of genes that are responsible
for skin color or morphology. The frequencies of these genes
can thus be very different from one population to another.
Most of the markers used by geneticists, by contrast, are neutral, in that none of the alleles studied provides any advantage
to the organism, and the frequency of the different alleles
reflects merely the origin and history of human populations.

This makes the study of such markers particularly interesting. The method used is the same as that for establishing an
evolutionary tree from molecular data: the more distinct the
allelic distribution between two human populations, the longer the two populations have been separated (even if this sep-
aration is not total and all genes continue to circulate within
the whole human population). By looking at the Y chromo-

some or at the mitochondrial DNA, it is even possible to discriminate between the movements of males and females.

These studies have not fundamentally changed what was already known about human migrations and the way we have populated the world. But they do show that today's population throughout the world is derived from a few thousand individuals living in Africa 100,000 to 200,000 years ago.[3] What was the nature of this genetic bottleneck that modern humans experienced? It is difficult to be sure, but whatever the answer, human beings are clearly genetically closer to each other than we previously thought. On the other hand, the results obtained from analysis of fossil DNA from Neanderthals shows that they were genetically very different from modern humans, despite the fact that the last Neanderthals were contemporaries of the first modern humans.[4]

The migration of modern humans from all over the world took place in several waves, leaving genetic traces that are still visible. Luigi Cavalli-Sforza has tried to link these genetic markers with key events in the history of humanity.[5] The development of agriculture from three or four centers which spread over the whole planet—the neolithic revolution—can be detected on maps showing the distribution of genetic markers. The technological revolution corresponding to the development of agriculture and animal husbandry led to a rapid growth in the population. Having become too numerous, the humans began to migrate, taking with them their new technology—and their genes.

To fully understand the history of humanity as told by our genes, we need on the one hand to avoid an error and on the other to accept a sad part of our past. The error that has to be avoided is to think that the people who migrated and colonized distant lands were in some way genetically superior to—

had better genes than—those they replaced. The genetic variation studied by geneticists is neutral. If the frequency of certain forms of gene has increased, this is simply because the populations that carried these forms grew more rapidly, and the new technology they employed made it possible to produce more abundant and safer food, thus making it possible to achieve a higher rate of reproduction.

The sad side to the story is that indigenous peoples who were subjected to these waves of migrating humans armed with more effective technology could have adopted these new technologies and advanced in turn. The migration of genes would have been attenuated. But the genetic analyses indicate that such a cultural exchange did not take place: the indigenous populations did not benefit from the new technologies, which remained linked to the genes of the populations that brought them. The absence of cultural and genetic exchanges that can be detected in the great population migrations following the neolithic revolution were repeated during later, less significant waves of migration.

Perhaps the exchanges were limited by the weak population density of humans at the time, far lower than for any modern population. The conclusions of all these studies should also be treated with caution: the proposed scenarios require an interpretation of the raw genetic data and the development of models that have yet to be fully validated. Recent studies using the Y chromosome have challenged previous models, tending to put the emphasis on culural exchange.[6]

Today, technological and cultural evolution has become so rapid that it affects the whole human population, leaving no time for gene frequencies to change. Furthermore, rates of reproduction are no longer connected to the technological capacities of different populations. That does not mean, however, that humanity's genetic evolution has come to an end. Cultural habits can have an influence on gene frequencies. The most convincing example is that of a gene involved in the absorption of the main sugar present in milk, lactose. Some people who carry a particular form of this gene do not make enough of the enzyme that makes it possible to use this sugar; when they consume too much milk, they develop nausea and diarrhea. The proportion of individuals bearing this form of the gene is far lower in populations performing animal husbandry for milk production. In other words, a cultural habit— the consumption of milk products—has created a strong selection pressure that, in a few generations, has profoundly modified a genetic characteristic of the human population.[7]

From this simple example it is possible to imagine other scenarios in which cultural preferences could change the frequency of gene forms—for example, by favoring or not the formation of certain couples of men and women. This is where science ends and science fiction begins. There is no proof that physical or psychological preferences could sufficiently influence the formation of couples to lead to a change in the forms of genes present in the human population. But if these preferences were strong enough to do so, would there not also be a risk that, in the near future, they would be expressed in another way, via the direct manipulation of genes? That is the subject of the next section.

EUGENICS

It would be difficult to close this discussion of gene function without tackling the fact that the knowledge we have gained is not neutral but can be used to modify humanity's genes. The projects and fantasies of the eugenicists need to be confronted with our new understanding of how genes do what they do.

Eugenic projects aimed at improving the quality of human stock predate the rise of genetics by many years.[8] Plato, in the ideal society outlined in his *Republic,* suggests that reproduction should be limited to the fittest. At the beginning of the nineteenth century, the French philosopher Georges Cabanis argued that humanity should take control of its reproduction and should practice on itself what it had been carrying out to such effect on animals and plants. Eugenics is an ideology whose roots go way back in the history of humankind.

However, the rise of genetics has profoundly altered the various eugenicist programs. On the one hand, it has finally revealed what makes humans—indeed all organisms—specific. The progress of molecular biology and the metaphors that have been proposed might even appear to have made the eugenic project more acceptable. Modifying the human genome would be simply a matter of correcting a badly written text, of removing a few spelling mistakes.

But above all, modern biology has developed tools that make it possible to change the genetic text. Simple, additive transgenesis, in which supplementary genetic material is introduced into an individual, makes it possible to enrich the genetic content of a given organism. Substitutional transgenesis by homologous recombination involves replacing an altered copy of a gene by a normal copy (and why not improve the normal copy?). In practice, the new genetic tools have rapidly given rise to a negative eugenics in which fetuses with genetic defects are eliminated at the early stages of prenatal development.[9] Diagnosis can be done even before the embryo is implanted in the uterus.

Discussions about proposals for genetically modifying hu-

mans are generally marked by two problems. First, they mix
up projects that are currently technically impossible but theoretically doable with those that are simply the product of a
profound ignorance of what genes actually do. Second, describing all such genetic projects as "eugenic," as their opponents currently do, does not help matters. Eugenics is not
linked inextricably to the manipulation of genes—a eugenic
project could easily exist without any understanding of genes,
and genes can be manipulated without a eugenic program.
The term "eugenic" should be reserved for those projects in
which an intervention in human reproduction is clearly aimed
at improving the human race.

Modifying the human genome is, in practice, technically
difficult because any such intervention has to be extremely effective. Imagine that two parents want to give their child a
new gene, or to eliminate a defective gene from its genome.
What they want above all is that such an intervention will
work. Current success rates of a few percent for additive
transgenesis, or of a few decimal points of one percent for
substitutional transgenesis, are sufficient for the food industry
in its search for a better cow, but they would be far too low for
a couple hoping for a normal, or "perfect," child.

Nevertheless, it would be wrong to allow the present
inefficiency of such techniques to debar the subject from discussion. Technical obstacles are often rapidly overcome. The
best example of this is the recently successful cloning of mammals.[10] Cloning involves the reproduction of an organism from
the DNA contained in any one of its cells. In theory, this is
possible because nearly all cells contain the same genetic information. In practice, it is impossible in animals because the
egg—the single cell from which the organisms develops and
which is the result of the fusion of a spermatozoid and an

oocyte—has, like the oocyte, a very special structure and properties which are necessary for early stages of development and for which there is no substitute.

On the other hand, it is possible to remove the nucleus from an oocyte—or an egg—and replace it with the nucleus of any old cell from an adult organism. This somewhat indirect method of cloning was first used on amphibians. The results were partly disappointing—an egg that has been reconstituted in this way can develop, but development is rarely complete, generally stopping after the first few cell divisions.

A few successes in mammals were announced at the beginning of the 1980s,[11] but these preliminary results were contested and the authors were accused of having faked them. Research groups, having failed to replicate the published results, concluded that cloning was theoretically impossible in mammals.[12] Two arguments were put forward to justify this conclusion. First, because the transcription of the mammalian embryo's genome begins very soon after fertilization, there is not enough time for a transplanted foreign nucleus to be "reprogrammed" by the oocyte. Second, the activity of parental and maternal genes differs at the beginning of development, due to the marking of genes during the maturation of the germ line cells—the process of genomic imprinting. This differential expression of the two copies of a given gene could not be reproduced by any old nucleus. However, these very good reasons for accepting that the cloning of mammals was impossible are in fact wrong.

At the beginning of the 1990s cows were cloned from early embryonic cells.[13] These results, which challenged the idea that cloning was impossible in mammals, went virtually unnoticed, as did the first cloning of sheep from cultured embryonic cells.[14] Only in 1997, when Dolly the sheep was cloned from a ewe's mammary gland cell, did these new techniques come into public awareness, being whipped up by a hurricane of media attention into a storm of controversy that still rages to this day.[15] And of course, it is the possible application of these techniques to human beings that has led to the current debate.

The debate on cloning has been made even more complex by the fact that clones introduce the possibility of producing embryonic stem cells.[16] These cells can be induced to differentiate *in vitro* to generate heart cells, neurons, or other cells that could then be used to replace corresponding deficient cells in human pathologies.[17] Such a therapeutic approach would allow physicians to reduce the number of human organ transplants and to avoid xenografts from animals. Because the cells used would be identical to the cells of the recipient organism, there would be no immune rejection of the transplanted cells. Such possibilities generated a huge—and still active—debate which is outside the scope of this book and which is mostly focused on the use of human embryos to generate these embryonic stem cells.

Two aspects of the cloning experiments have not been widely discussed but are particularly interesting for the questions of gene action and transgenesis. First, what are the similarities or differences between individuals produced from the cells of a given organism? Clones will never be as similar as identical twins because they do not share the same mitochondrial DNA, maternal oocyte, or intrauterine environment, as identical twins do. Any human clones would be distinct individuals, even if they more closely resembled one another than do people in the general population (identical twins excepted). Comparing animal clones produced from cells from the same organism would be an excellent way of estimating the role of nuclear genes separate from the influence of the environment and mitochondrial genes—something that is not possible in twin studies.

Recent cloning experiments are also particularly important for they make it possible to carry out well-controlled additive and substitutional transgenesis.[18] All that is necessary is to start from embryonic cells, modify their genes, check that the modification has worked (this was not previously possible), and then insert the nuclei of such cells into oocytes, thus obtaining viable embryos. This was how scientists created Polly, the lamb whose milk contained a human blood coagulation factor that is missing in hemophiliacs.[19] The technique will have to be made more efficient, but it seems there is no theoretical reason that it cannot be used for human transgenesis.

This immediately raises the question of the objective of any change to the human genome. What does the study of gene function outlined here tell us about what can and what cannot be altered? What is the relation between these possibilities and the eugenic projects that marked the first half of the twentieth century? At the time, the aim of such policies was to avoid the spread of "bad" genes, made possible by the weakening of natural selection on our species. Although forced sterilization was practiced in various countries up until the end of the 1960s, in general, eugenic policies were abandoned after the Second World War. The horror of eugenic policies in Nazi Germany contributed a great deal to this change. But scientific reasons also played a part.[20]

The discovery of the complexity of the genetic control of various defects and the important role of the environment in many syndromes—in particular psychological problems—that had previously been given a purely genetic etiology made eugenic policies seem simplistic. Furthermore, population geneticists showed early on that any policy aimed at preventing the reproduction of affected individuals (who generally carry two altered copies of the gene in question) would only very slowly reduce the frequency of these altered genes in the population and would therefore have no immediate visible effect on the observed frequency of the defects, since most of the altered copies of genes are transmitted from generation to generation by individuals with no outward signs of the defective gene.

Finally, some studies of human genetics suggested that in and of itself a gene is not good or bad—that depends on its genetic and environmental context. Thus the genetic form that is responsible for sickle-cell anemia provides the individuals who carry it with an increased resistance to malaria, which in some parts of the world has been life-saving—and that fact accounts for the survival of the allele in the population. Similarly, it has recently been suggested that in a heterozygous state, the allele responsible for cystic fibrosis might protect carriers from typhoid fever.[21]

Where do such arguments stand up against our present understanding of genes? The idea of the "bad" gene, which was ridiculed by the opponents of any eugenic policy, is not completely absurd if it is applied to particular forms of genes. It is difficult to see how the mutations that produce Huntington disease, which leads to death by neurodegeneration at around 50 years of age, could conceivably be considered good or favorable, even in a very special genetic or environmental context. If this disease has not been eliminated by natural selection, it is no doubt because, like many other diseases of old age, it affects individuals past reproductive age. To imagine that there might be something good about this mutated form would be to accept that the life of a human being has no value once they have lost the possibility of transmitting their genes to the next generation. Hardly a civilized view!

If a simple technique could eliminate the allele responsible for Huntington disease from the human genome overnight, it is unlikely that many tears would be shed over this impoverishment of human genetic diversity. Similarly, given that many infectious diseases have been conquered or are in retreat, the benefit that mutations such as that which causes cystic fibrosis might provide to heterozygote carriers no longer exists. All that remains is the catastrophic effect of the mutation in affected homozygotes.

There are two kinds of policies that aim at modifying the human genome. There are those that, from a therapeutic point of view, are intended to change the genetic content of cells, sometimes in order to introduce a function that was absent or inefficient, often to introduce killer genes to eliminate cancer cells. Genes that are manipulated as part of such somatic gene therapy are not passed on—this has to be checked very carefully—and thus the human genome is not altered. The genes involved generally code for well-known proteins with a single function. These projects, known as gene therapy, do not need to be discussed much. They are simply a new therapeutic technique. Giving an injection in order to introduce an absent blood factor or putting into certain cells genes that make it possible to synthesize such a factor are two equivalent therapeutic interventions; the second merely acts upstream, prior to the synthesis of the factors in question. The second approach should not be criticized because it uses genes. The only criteria that should guide the development of such new therapeutic practices are those of harmlessness and efficacy: somatic gene therapy is justified only if it is better for the patients than other methods, in terms of the benefits it brings, given the problems associated with any treatment.

The real debate centers around the modification of genes in the egg or in the cells that produce eggs and sperm—a change that would be transmitted to an individual's offspring. For the moment, such experiments are banned for the simple reason that this treatment technique is currently inefficient and thus without any clear benefit. At present, all we can do in the case of a genetic disease where affected individuals have altered copies of the gene in question is to propose to the parents at risk that they undergo prenatal diagnosis so that, if they so desire, the affected fetuses can be aborted.

Given these current practices, what would be our interest in pursuing germ-line gene therapy? It could be applied if the parents refused abortion or, in the case of a diagnosis prior to implantation, if they refused to consider destroying the embryo but still wanted to have a child. Since abortion, even "convenience" abortion, is widely accepted today, it seems unlikely that, in a near future, such motives will be sufficient to encourage the development of germ-line gene therapy.

However, the situation might evolve. The number of genetic diseases for which a prenatal diagnosis is available is increasing continuously. Soon it will be possible to diagnose slight anomalies, and most doctors will refuse to perform abortions for minor defects. But at the same time, most parents will not want their children to be affected by such defects. In the absence of any alternative therapy, the only solution will be the development of a germ-line gene therapy.

There is also a second justification: the desire not to have to repeat, from generation to generation, the same diagnostic tests and therapeutic interventions—the wish to eradicate the problem once and for all. In most cases, however, such motivations come up against the problem that most of the people who pass the "bad" gene down the generations are themselves healthy and have healthy offspring. There would be no valid reasons to propose a complex, risky therapy to healthy families. If germ-line gene therapy was applied only to affected parents, it would be utterly ineffective in reducing the number of affected genes in the population as a whole, and thus the number of prenatal or preimplantation diagnostic tests that would be required would not decrease. Only in the case of dominant genetic diseases, such as Huntington's, where everyone carrying even one copy of the defective allele exhibits the disease, would germ-line gene therapy lead to a decrease in the number of affected individuals over a few generations. But in this case, aborting embryos that carry the mutation would lead to the same result in a much simpler way.

Presently, germ-line gene therapy is banned not just because of the practical difficulties in applying it to human beings but because of a widespread fear that it would impoverish the human gene pool by reducing its diversity, in however limited a fashion, and, more fundamentally, because of an opposition to the idea of tinkering with the human genome at all.

The reasoning behind this rejection of germ-line gene therapy merits thoughtful discussion. The idea that genetic diversity is so good in itself that any reduction in diversity, however limited, should be opposed is highly debatable. Furthermore, it would take several centuries of work by all the biologists on the planet to reduce genetic diversity to the same extent that the various wars and waves of colonization have done throughout our history, including the twentieth century. When the activity of the first *Homo sapiens sapiens* led, directly or indirectly, to the disappearance of Neanderthals, this may have deprived the humanoid genome of forms of genes that could perhaps have opened new evolutionary possibilities.

The second reason put forward to justify such a ban is that by changing our genome we would be violating some sacred taboo, or crossing a boundary that would launch us on a voyage of no return. Once again, it should be noted that ever since humans have reproduced (which is, of course, always) we have manipulated our genes and varied allelic frequencies, increasing some and decreasing others. Why should genes be more sacred than the people bearing them? Opponents and partisans of genetic manipulation often share the same fascination with genes and their power, and the same ignorance of their real function and place in the organization of life.

The only serious reason why we should ban genetic manipulation of the human germ line is our ignorance, or more precisely the weakness of our current knowledge and, on the other hand, the complexity of life that has been already illuminated by our limited understanding. Biologists and geneticists have often wanted to go too fast, to apply their knowledge to the well-being of humanity without realizing the limits of this knowledge and the problems that could arise from hasty action.[22] Informed by such errors, we should wait before altering the human genome. Despite the real possibilities described here, the underlying techniques available to geneticists are still inefficient and uncontrolled.

Our understanding is even less certain. Throughout this book, I have tried to give a clear description of gene action, or more precisely of protein—the product of gene—action. I have emphasized the paradox that is involved in according a fundamental role to genes in even the most complex characteristics of organisms, and the impossibility of stating precisely what this role is. This impossibility flows from the phenomena of redundancy and compensation, the self-organization of the components of life in networks, and the assembly of these elementary components in hierarchical structures.[23] How can we hope to change such a complex organization without destabilizing it, without provoking devastating secondary effects? We are many, many years away from being able to modify the human genome and fully understand what we are doing, realizing in advance the consequences of every change. But we can be sure that such a time will come.

Only then will the question of the aim of such changes be really posed. To prevent genetic diseases? We have seen the problems associated with this. To change humanity? If so, how, and in what direction?

NOTES

INTRODUCTION

1. Sarkar, S. (1998), *Genetics and reductionism* (Cambridge, UK: Cambridge University Press). Beurton, P. J., Falk, R., and Rheinberger H. J., eds. (2000), *The concept of the gene in development and evolution* (New York: Cambridge University Press).

2. Jacquard, A. (1984), *In praise of difference: Genetics and human affairs* (New York: Columbia University Press).

3. Genes also control the synthesis of some RNAs that are not translated into proteins but which play structural or other functions.

4. Deamer, D. W., and Fleischaker, G. R. (1994), *Origins of life: The central concepts* (Boston: John and Bartlett Publishers); Orgel, L. E. (1998), "The origin of life—a review of facts and speculations," *TIBS* 23: 491–495.

5. Plasterk, R. H. A. (1999), "Hershey heaven and *Caenorhabditis elegans*," *Nature genetics* 21: 63–64.

6. Ostolaza, J. F., and Bergareche, A. M. (1992), *Vida artificial* (Madrid: Eudema).

7. Flint, J. (1996), "The genetics of crime: Irresponsible science?" *TIG* 12: 240.

8. Hamer, D. H. (1996), "The heritability of happiness," *Nature genetics* 14: 125–126.

9. Rose, S. (1998), *Lifelines: Biology beyond determinism* (New York: Oxford University Press).

10. Lewontin, R. C., Rose, R., and Kamin, L. J. (1984), *Not in our genes* (New York: Pantheon Books).

11. Hamer, D., and Copeland, P. (1998), *Living with our genes* (New York: Anchor Books, Doubleday), 11.

1. THE CONCEPT OF THE GENE

1. Carlson, E. A. (1966), *The gene: A critical history* (Philadelphia: Saunders).

2. Kevles, D. (1985), *In the name of eugenics* (New York: Alfred A. Knopf).

3. Allen, G. E. (1978), *Thomas Hunt Morgan: The man and his science* (Princeton, NJ: Princeton University Press).

4. Morgan, T. H., Sturtevant, A. H., Muller, H. J., and Bridges, C. B. (1915), *The mechanism of Mendelian heredity* (New York: Henry Holt and Co.).

5. Allen, G. E., *Thomas Hunt Morgan*, 263–278.

6. Allen, G. E. (1974), "Opposition to the Mendelian chromosome theory: The physiological and developmental genetics of Richard Goldschmidt," *J. hist. biol.* 7: 49–92.

7. East, E. M. (1912), "The Mendelian notation as a description of physiological facts," *Am. natur.* 46: 633–655.

8. Olby, R. (1974), *The path to the double helix* (London: MacMillan; new ed. 1994, Dover Publications).

9. Judson, H. F. (1996), *The eighth day of creation: The makers of the revolution in biology,* expanded ed. (Cold Spring Harbor, NY: Cold Spring Harbor Laboratory Press).

10. Gayon, J. (1998), *Darwinism's struggle for survival: Heredity and the hypothesis of natural selection* (Cambridge, UK: Cambridge University Press).

11. Kimura, M. (1983), *The neutral theory of molecular evolution* (Cambridge, UK: Cambridge University Press).

12. Gould, S. J. (1989), *Wonderful life: The burgess shale and the nature of history* (New York: W. W. Norton).

13. Williams, G. C. (1966), *Adaptation and natural selection* (Princeton, NJ: Princeton University Press).

14. Jacob, F. (1982), *The possible and the actual* (Seattle, WA: University of Washington Press).

15. Dennett, D. L. (1995), *Darwin's dangerous idea* (New York: Simon and Schuster).

16. Keller, E. F. (1995), *Refiguring life: Metaphors of twentieth century biology* (New York: Columbia University Press). See also Keller, E. F. (2000), *The century of the gene* (Cambridge: Harvard University Press).

17. Kay, L. E. (1995), "Who wrote the book of life? Information and the transformation of molecular biology, 1945–55," *Science in context* 8: 609–634; Kay, L. E. (1997), "Cybernetics, information, life: The emergence of scriptural representations of heredity," *Configurations* 5: 23–91; Kay, L. E. (2000), *Who wrote the book of life: A history of the genetic code (writing science)* (Stanford, CA: Stanford University Press).

18. Jacob, F. (1973), *The logic of life* (Princeton, NJ: Princeton University Press).

19. Atlan, H. (1979), *Entre le cristal et la fumée: essai sur l'organisation du vivant* (Paris: Le Seuil), p. 56.

2. Can We Get Rid of the Gene Concept?

1. It should not be forgotten that some genes do not code for proteins but for RNAs, either ribosomal RNAs that make up ribosomes (cellular structures upon which protein synthesis takes place), transfer RNAs that "carry" amino acids during protein synthesis, or a number of small RNAs that carry out various functions in the cell. These RNAs are very probably the visible traces of the "RNA world" that preceded the living world we know today.

2. Portin, P. (1993), "The concept of the gene: Short history and present status," *Quart. rev. biol.* 68: 173–223; Beurton, P. J., Lefèvre, W., and Rheinberger, H. J. (1995), "Gene concepts and evolution," *Max-Planck-Institut für Wissenschaftsgeschichte* n 18.

3. For a definition of loose concepts, see Elkana, Y. (1970), "Helmoltz's 'kraft': An illustration of concept in flux," *Hist. stud. phys. sci.* 2: 263–298. See also Löwy, I. (1992), "The strength of loose concepts—boundary concepts, federative experimental strategies and disciplinary growth: The case of immunology," *Hist. sci.* 30: 371–396. On the vagueness of the gene as a concept, see Keller, E.F. (2000), *The century of the gene* (Cambridge: Harvard University Press).

4. Scott, J. (1995), "A place in the world for RNA editing," *Cell* 81: 833–836.

5. Beurton, P. J., et al., (1995), "Gene concepts and evolution," *Max-Planck-Institut für Wissenschaftsgeschichte* n 18.

6. Goldschmidt, R. B. (1946), "Position effect and the theory of the corpuscular gene," *Experientia* 2: 197–203 and 250–256.

7. The *C. elegans* sequencing consortium (1998), "Genome sequence of the nematode *C. elegans*: A platform for investigating biology," *Science* 282: 2012–2018.

8. Brenner, S., et al. (1993), "Characterization of the pufferfish (*Fugu*) genome as a compact model vertebrate genome," *Nature* 366: 265–268.

9. The *C. elegans* sequencing consortium (1998), "Genome sequence of the nematode *C. elegans*," *Science* 282: 2012–2018.

10. Dunham, I., et al. (1999), "The DNA sequence of human chromosome 22," *Nature* 402: 489–495; Hattori, M., et al. (2000), "The DNA sequence of human chromosome 21," *Nature* 405: 311–319. Adams, M.D., et al. (2000), "The genome sequence of *Drosophila Melanogaster*," *Science* 287: 2185–2195.

11. Mewes, H. W., et al. (1997), "Overview of the yeast genome," *Nature* 387 Suppl.: 7–8.

12. Gabor Miklos, G. L., and Rubin, G. M. (1996), "The role of the genome project in determining gene function: Insights from model organisms," *Cell* 86: 521–529.

13. Wolfe, K. H., and Shields, D. C. (1997), "Molecular evidence for an ancient duplication of the entire yeast genome," *Nature* 387: 708–713.

14. O'Brien, S. J. (1973), "On estimating functional gene number in eukaryotes," *Nature new biology* 242: 52–54.

15. I have tried to avoid using "epigenetic" to describe loose and indirect genetic control. This word means different things to different people and its use thus tends to confuse more than it clarifies. Chernoff, Y. O. (1999), "A black spot of modern biology," *TIG* 15: 336.

16. Sapp, J. (1987), *Beyond the gene: Cytoplasmic inheritance and the struggle for authority in genetics* (New York: Oxford University Press). This question was studied with particular intensity in Germany: Harwood, J. (1993), *Style of scientific thought: The German genetics community, 1910–1933* (Chicago: The University of Chicago Press).

17. Warren, G., and Wickner, W. (1996), "Organelle inheritance" *Cell* 84: 395–400.

18. Hartl, F. U. (1996), "Molecular chaperones in cellular protein folding," *Nature* 381: 571–580.

19. Prusiner, S. B. (1982), "Novel proteinaceous infectious particles cause scrapie," *Science* 216: 136–144; Prusiner, S. B. (1991), "Molecular biology of prion diseases," *Science* 252: 1515–1522; Keyes, M. E. (1999), "The prion challenge to the 'central dogma' of molecular biology, 1965–1991," *Stud. hist. phil. biol. & biomed. sci.* 30: 1–19 and 181–218.

20. Horwich, A. L., and Weissman, J. S. (1997), "Deadly conformations— protein misfolding in prion disease," *Cell* 89: 499–510.

21. Lindquist, S. (1997), "Mad cows meet psi-chotic yeast: The expansion of the prion hypothesis," *Cell* 89: 495–498; Sondheimer, N., and Lindquist, S. (2000), "Rnq1: an epigenetic modifier of protein function in yeast, "*Molecular Cell* 5: 163–172. Coustou, V., Deleu, C., Saupe, S., and Begueret, J. (1997), "The protein product of the *het-s* heterokaryon incompatibility gene of the fungus *Podospora anserina* behaves as a prion analog," *Proc. natl. acad. sci. USA* 94: 9773–9778.

3. INVESTIGATING WHAT GENES REALLY DO

1. Capecchi, M. R. (1989), "Altering the genome by homologous recombination," *Science* 244: 1288–1292.

2. Collins, F. S. (1995), "Positional cloning moves from perditional to traditional," *Nature genetics* 9: 347–350.

4. GENES THAT CAUSE DISEASES

1. Neel, J. V. (1949), "The inheritance of sickle cell anemia," *Science* 110: 64–66.

2. Pauling, L., Itano, H. A., Singer, S. J., and Wells, I. C. (1949), "Sickle cell anemia, a molecular disease," *Science* 110: 543–548.

3. Ingram, V. M. (1957), "Gene mutations in human haemoglobin: The chemical difference between normal and sickle cell haemoglobin," *Nature* 180: 326–328.

4. Editorial (1996), "What are genes *for?*" *Nature genetics* 14: 235–236; Judson, H. F. (1996), *The eighth day of creation: Makers of the revolu-*

tion in biology, expanded ed. (Cold Spring Harbor, NY: Cold Spring Harbor Laboratory Press), p. 607.

It would be wrong of researchers to suggest that the popular media are responsible for oversimplifying their results when the main culprits are in fact the specialists themselves. For example, on 5 February 1999, the excellent journal *Cell*, one of the most prestigious scientific publications in the world, had as its cover headline "Potassium channel causes deafness." Although the title of the article corrected this oversimplification by explaining that a member of the potassium channel family was found to be mutated in cases of dominant deafness, by then the damage had been done. Kubisch, C., et al. (1999), "KCNQ4, a novel potassium channel expressed in sensory outer hair cells, is mutated in dominant deafness," *Cell* 96: 437–446.

5. Labie, D., and Elion, J. (1996), "Modulation polygénique des maladies monogéniques: L'exemple de la drépanocytose," *Médecine/sciences* 12: 341–349.

6. Penrose, L. S. (1946), "Phenylketonuria: A problem in eugenics," *Lancet* 1: 949–953; Scriver, C. R., and Waters, P. J. (1999), "Monogenic traits are not simple: Lessons from phenylketonuria," *TIG* 15: 267–272.

7. Férec, C., Mercier, B., and Audrézet, M.-P. (1994), "Les mutations de la mucoviscidose: Du génotype au phénotype," *Médecine/sciences* 10: 631–639.

8. Rigot, J. M., et al. (1991), "Cystic fibrosis and congenital absence of the vas deferens," *New England journal of medicine* 325: 64–65.

9. Mulvihill, J. J. (1995), "Craniofacial syndromes: No such thing as a single gene disease," *Nature genetics* 9: 101–103.

10. Kauffman, S. A. (1993), *The origins of order: Self organization and selection in evolution* (New York: Oxford University Press).

11. Monaco, A. P. (1996), "Dissecting Williams syndrome," *Current biology* 6: 1396–1398.

12. Frangiskakis, J. M., et al. (1996), "*LIM-kinase 1* hemizygosity implicated in impaired visuospatial constructive cognition," *Cell* 86: 59–69.

13. Tassabehji, M., et al. (1999), "Williams syndrome: use of chromosomal microdeletions as a tool to dissect cognitive and physical phenotypes," *American journal of human genetics* 64: 118–125.

14. Nasir, J., et al. (1995), "Targeted disruption of the Huntington's disease gene results in embryonic lethality and behavioral and morphological changes in heterozygotes," *Cell* 81: 811–823.

15. Mangiarini, L., et al. (1996), "Exon 1 of the *HD* gene with an expanded CAG repeat is sufficient to cause a progressive neurological phenotype in transgenic mice," *Cell* 87: 493–506.

16. Yamamoto, A., Lucas, J. J., and Hen, R. (2000), "Reversal of neuropathology and motor dysfunction in a conditional model of Huntington's disease," *Cell* 101: 57–66.

17. Davies, S. W., et al. (1997), "Formation of neuronal intranuclear inclusions underlies the neurological dysfunction in mice transgenic for the HD mutation," *Cell* 90: 537–548; Scherzinger, E., et al. (1997), "Huntingtin-encoded polyglutamine expansions form amyloid-like protein aggregates in vitro and in vivo," *Cell* 90: 549–558; Martindale, D., et al. (1998), "Length of huntingtin and its polyglutamine tract influences localization and frequency of intracellular aggregates," *Nature genetics* 18: 150–154; Saudou, F., Finkbeiner, S., Devys, D., and Greenberg, M. E. (1998), "Huntingtin acts in the nucleus to induce apoptosis but death does not correlate with the formation of intranuclear inclusions," *Cell* 95: 55–66.

18. Van Dellen, A., Blakemore, C., Deacon, R., York, D., and Hannan, A. J. (2000), "Delaying the onset of Huntington's in mice," *Nature* 404: 721–722.

19. Georgiou, N., et al. (1999)," Differential clinical and motor control function in a pair of monozygotic twins with Huntington's disease," *Movement disorders* 14: 320–325.

20. Tanzi, R. E., and Parson, A. B. (2000), *Decoding darkness: the search for the genetic causes of Alzheimer's disease* (Boston: Perseus). Mark, R. J., et al. (1997), "Amyloid β-peptide impairs glucose transport in hippocampal and cortical neurons: Involvement of membrane lipid peroxidation," *J. neurosci.* 17: 1046–1054.

21. De Chadarevian, S. (1998), "Of worms and programs: *Caenorhabditis elegans* and the study of development," *Stud. hist. phil. biol. & biomed. sci.* 29: 81–105.

22. Levitan, D., and Greenwald, I. (1995), "Facilitation of *lin-12*-mediated signaling by *sel-12*, a *Caenorhabditis elegans S182* Alzheimer's disease gene," *Nature* 377: 351–354.

23. Duff, K., et al. (1996), "Increased amyloid-β42(43) in brains of mice expressing mutant presenilin 1," *Nature* 383: 710–713.

24. Wolfe, M. S., et al. (1999), "Two transmembrane aspartates in presenilin-1 required for presenilin endoproteolysis and γ-secretase activity," *Nature* 398: 513–517. Li, Y.-M., et al. (2000) "Photoactivated γ-secretase inhibitors directed to the active site covalently label presenilin 1," *Nature* 405: 689–694.

25. Holcomb, L., et al. (1998), "Accelerated Alzheimer-type phenotype in transgenic mice carrying both mutant *amyloid precursor protein* and *presenilin 1* transgenes," *Nature medicine* 4: 97–100.

26. Gervais, F. G., et al. (1999), "Involvement of caspases in proteolytic cleavage of Alzheimer's amyloid-β precursor protein and amyloidogenic Aβ peptide formation," *Cell* 97: 395–406.

27. Meyer, M. R., et al. (1998), "*APOE* genotype predicts when—not whether—one is predisposed to develop Alzheimer disease," *Nature genetics* 19: 321–322; Blacker, D., et al. (1998), "Alpha-2 macroglobulin is genetically associated with Alzheimer disease," *Nature genetics* 19: 357–360.

28. Wolf, U. (1995), "The genetic contribution to the phenotype," *Hum. genet.* 95: 127–148.

5. KNOCKOUT SURPRISES

1. Other techniques, such as the use of antisense RNA or RNA interference, also give biologists the means to control gene expression. We will focus on gene knockouts, which provided the richest results.

2. Brandon, E. P., Idzerda, R. L., and McKnight, G. S. (1995), "Targeting the mouse genome: A compendium of knockouts," *Current biology* 5: 625–634, 758–765, and 873–881; see also the mouse knockout database: http://biomednet.com/cgi-bim/mko/mkohome.pl

3. Picciotto, M. R., et al. (1995), "Abnormal avoidance learning in mice lacking functional high-affinity nicotine receptor in the brain," *Nature* 374: 65–67.

4. De la Pompa, J. L., et al. (1998), "Role of the NF-ATc transcription factor in morphogenesis of cardiac valves and septum," *Nature* 392: 182–186; Ranger, A. M., et al. (1998), "The transcription factor NF-ATc is essential for cardiac valve formation," *Nature* 392: 186–190.

5. Fernandez-Salguero, P., et al. (1995), "Immune system impairment and hepatic fibrosis in mice lacking the dioxin-binding Ah receptor," *Science* 268: 722–726.

6. DeWitt, D., and Smith, W. L. (1995), "Yes, but do they still get headaches?" *Cell* 83: 345–348.

7. Büeler, H., et al. (1992), "Normal development and behavior of mice lacking the neuronal cell-surface PrP protein," *Nature* 356: 577–582; Tobler, I., et al. (1996), "Altered circadian activity rhythms and sleep in mice devoid of prion protein," *Nature* 380: 639–642.

8. Hooper, M., et al. (1987), "HPRT-deficient (Lesch-Nyhan) mouse embryos derived from germline colonization by cultured cells," *Nature* 326: 292–295; Kuehn, M. R., Bradley, A., Robertson, E. J., and Evans, M. J. (1987), "A potential animal model for Lesch-Nyhan syndrome through introduction of HPRT mutations into mice," *Nature* 326: 295–298.

   This hypothesis was confirmed: symptoms similar to those observed in the human disease were obtained in the mutated mice by inhibiting the activity of another enzyme involved in the synthesis of the same DNA and RNA components. Wu, C.-L., and Melton, D. W. (1993), "Production of a model for Lesch-Nyhan syndrome in hypoxanthine phosphoribosyltransferase-deficient mice," *Nature genet.* 3: 235–240.

9. Baribault, H., Price, J., Miyai, K., and Oshima, R. G. (1993), "Midgestational lethality in mice lacking keratin 8," *Genes dev.* 7: 1191–1202; Baribault, H., Penner, J., Iozzo, R. V., and Wilson-Heiner, M. (1994), "Colorectal hyperplasia and inflammation in keratin 8-deficient FVB/N mice," *Genes dev.* 8: 2964–2973; Threadgill, D. W., et al. (1995), "Targeted disruption of mouse EGF receptor: Effect of genetic background on mutant phenotype," *Science* 269: 230–234; Sibilia, M., and Wagner, E. F. (1995), "Strain-dependent epithelial defects in mice lacking the EGF receptor," *Science* 269: 234–238.

10. Morange, M. (1998), *A history of molecular biology* (Cambridge, MA: Harvard University Press).

11. Larue, L., Ohsugi, M., Hirchenhain, J., and Kemler, R. (1994), "E-cadherin null mutant embryos fail to form a trophectoderm epithelium," *Proc. natl. acad. sci. USA* 91: 8263–8267.

12. Saga, Y., et al. (1992), "Mice develop normally without tenascin," *Genes dev.* 6: 1821–1831; Crossin, K. L. (1994), "Functional role of cytotactin/

tenascin in morphogenesis: A modest proposal," *Perspectives on developmental neurobiology* 2: 21–32.

13. Pickett, F. B., and Meeks-Wagner, D. R. (1995), "Seeing double: Appreciating genetic redundancy," *The plant cell* 7: 1347–1356.

14. Thomas, J. H. (1993), "Thinking about genetic redundancy," *TIG* 9: 395–399; Nowak, M. A., Boerlijst, M. C., Cooke, J., and Maynard Smith, J. (1997), "Evolution of genetic redundancy," *Nature* 388: 167–171.

15. Fuchs, E., and Cleveland, D. W. (1998), "A structural scaffolding of intermediate filaments in health and disease," *Science* 279: 514–519.

16. Colucci-Guyon, E., et al. (1994), "Mice lacking vimentin develop and reproduce without an obvious phenotype," *Cell* 79: 679–694.

17. Gomi, H., et al. (1995), "Mice devoid of the glial fibrillary acidic protein develop normally and are susceptible to scrapie prions," *Neuron* 14: 29–41.

18. Shibuki, K., et al. (1996), "Deficient cerebellar long-term depression, impaired eyeblink conditioning, and normal motor coordination in GFAP mutant mice," *Neuron* 16: 587–599; McCall, M. A., et al. (1996), "Targeted deletion in astrocyte intermediate filament (*Gfap*) alters neuronal physiology," *Proc. natl. acad. sci. USA* 93: 6361–6366; Liedtke, W., et al. (1996), "GFAP is necessary for the integrity of CNS white matter architecture and long-term maintenance of myelination," *Neuron* 17: 607–615.

19. Raeber, A. J., et al. (1997), "Astrocyte-specific expression of hamster prion protein (PrP) renders PrP knockout mice susceptible to hamster scrapie," *EMBO j.* 16: 6057–6065.

20. Henkemeyer, M., et al. (1995), "Vascular system defects and neuronal apoptosis in mice lacking Ras GTPase-activating protein," *Nature* 377: 695–701.

21. Soriano, P., Montgomery, C., Geske, R., and Bradley, A. (1991), "Targeted disruption of the c-*src* proto-oncogene leads to osteopetrosis in mice," *Cell* 64: 693–702.

22. Imamoto, A., and Soriano, P. (1993), "Disruption of the *csk* gene, encoding a negative regulator of Src family tyrosine kinases, leads to neural tube defects and embryonic lethality in mice," *Cell* 73: 1117–1124.

23. Nada, S., et al. (1993), "Constitutive activation of Src family kinases in mouse embryos that lack Csk," *Cell* 73: 1125–1135.

24. Hilberg, F., Aguzzi, A., Howells, N., and Wagner, E. F.(1993), "c-Jun is essential for normal mouse development and hepatogenesis," *Nature* 365: 179–181.

25. Johnson, R. S., Spiegelman, B. M., and Papaioannou, V. (1992), "Pleiotropic effects of a null mutation in the c-*fos* proto-oncogene," *Cell* 71: 577–586; Wang, Z.-Q., et al. (1992), "Bone and haematopoietic defects in mice lacking c-*fos*," *Nature* 360: 741–745.

26. Gruda, M. C., et al. (1996), "Expression of FosB during mouse development: Normal development of FosB knockout mice," *Oncogene* 12: 2177–2185.

27. Brown, J. R., et al. (1996), "A defect in nurturing in mice lacking the immediate early gene *fosB*," *Cell* 86: 297–309.

6. MOLECULES TO MIND

1. Goelet, P., Castellucci, V. F., Schacher, S., and Kandel, E. R. (1986), "The long and the short of long-term memory—a molecular framework," *Nature* 322: 419–422. McGaugh, J. L. (2000), "Memory—a century of consolidation," *Science* 287: 248–251.

2. Morris, R. G. M., Garrud, P., Rawlins, J. N. P., and O'Keefe, J. (1982), "Place navigation impaired in rats with hippocampal lesions," *Nature* 297: 681–683.

3. Bliss, T. V. P., and Collingridge, G. L. (1993), "A synaptic model of memory: Long-term potentiation in the hippocampus," *Nature* 361: 31–39.

4. Quinn, W. G., Harris, W. A., and Benzer, S. (1974), "Conditioned behavior in *Drosophila melanogaster*," *Proc. natl. acad. sci. USA* 71: 708–712; Weiner, J. (1999), *Time, love, memory. A great biologist and his quest for the origins of behavior* (New York: Alfred A. Knopf).

5. Belvin, M. P., and Yin, J. C. P. (1997), "*Drosophila* learning and memory: Recent progress and new approaches," *BioEssays* 19: 1083–1089.

6. Griffith, L. C., et al. (1993), "Inhibition of calcium/calmodulin-dependent protein kinase in Drosophila disrupts behavioral plasticity," *Neuron* 10: 501–509.

7. Cremer, H., et al. (1994), "Inactivation of the N-CAM gene in mice results in size reduction of the olfactory bulb and deficits in spatial learning," *Nature* 367: 455–459.

198 NOTES TO PAGES 86–90

8. Wu, Z.-L., et al. (1995), "Altered behavior and long-term potentiation in type I adenylyl cyclase mutant mice," *Proc. natl. acad. sci. USA* 92: 220–224.

9. Bourtchuladze, R., et al. (1994), "Deficient long-term memory in mice with a targeted mutation in the cAMP-responsive element-binding protein," *Cell* 79: 59–68.

10. Grant, S. G. N., et al. (1992), "Impaired long-term potentiation, spatial learning, and hippocampal development in *fyn* mutant mice," *Science* 258: 1903–1910.

11. Silva, A. J., Stevens, C. F., Tonegawa, S., and Wang, Y. (1992), "Deficient hippocampal long-term potentiation in α-calcium-calmodulin kinase II mutant mice," *Science* 257: 201–206.

12. Silva, A. J., Paylor, R., Wehner, J. M., and Tonegawa, S. (1992), "Impaired spatial learning in α-calcium-calmodulin kinase II mutant mice," *Science* 257: 206–211.

13. Morris, R. G. M., Anderson, E., Lynch, G. S., and Baudry, M. (1986), "Selective impairment of learning and blockade of long-term potentiation by an N-methyl-D-aspartate receptor antagonist, AP5," *Nature* 319: 774–776.

14. Tsien, J. Z., Huerta, P. T., and Tonegawa, S. (1996), "The essential role of hippocampal CA1 NMDA receptor-dependent synaptic plasticity in spatial memory," *Cell* 87: 1327–1338.

15. McHugh, T. J., et al. (1996), "Impaired hippocampal representation of space in CA1-specific NMDAR1 knockout mice," *Cell* 87: 1339–1349.

16. Tang, Y.-P., et al. (1999), "Genetic enhancement of learning and memory in mice," *Nature* 401: 63–69.

17. Tang et al., ibid., ix.

18. Palmiter, R. D., et al. (1982), "Dramatic growth of mice that develop from eggs microinjected with metallothionein-growth hormone fusion genes," *Nature* 300: 611–615.

19. Stevens, C. F. (1996), "Spatial learning and memory: The beginning of a dream," *Cell* 87: 1147–1148.

20. Lam, H.-M., et al. (1998), "Glutamate-receptor genes in plants," *Nature* 396: 125–126.

21. Schwartz, J. H. (1993), "Cognitive kinases," *Proc. natl. acad. sci. USA* 90: 8310–8313.

7. GENES CONTROLLING LIFE AND DEATH

1. Bishop, J. M. (1995), "Cancer: The rise of the genetic paradigm," *Genes dev.* 9: 1309–1315; Morange, M. (1993), "The discovery of cellular oncogenes," *Hist. phil. life sci.* 15: 45–58; Fujimura, J. H. (1996), *Crafting science: A sociohistory of the quest for the genetics of cancer* (Cambridge, MA: Harvard University Press).

2. Morange, M. (1997), "From the regulatory vision of cancer to the oncogene paradigm, 1975–1985," *J. hist. biol.* 30: 1–29.

3. Hanahan, D., and Weinberg, R. A. (2000), "The hallmarks of cancer," *Cell* 100: 57–70.

4. Hahn, W. C., et al. (1999), "Creation of human tumor cells with defined genetic elements," *Nature* 400: 464–468.

5. See for instance: Markert, C. L. (1968), "Neoplasia: A disease of cell differentiation," *Cancer res.* 28: 1908–1914.

6. Morange, M. (1996), "Construction of the developmental gene concept. The crucial years: 1960–1980," *Biol. zent. bl.* 115: 132–138.

7. King, M.-C., and Wilson, A. (1975), "Evolution at two levels in humans and chimpanzees," *Science* 188: 107–116.

8. Gould, S. J. (1977), *Ontogeny and phylogeny* (Cambridge, MA: Harvard University Press).

9. Raff, R. A., and Kaufman, T. C. (1983), *Embryos, genes and evolution* (London: MacMillan).

10. Goldschmidt, R. (1940), *The material basis of evolution* (New Haven, CT: Yale University Press), reedited in 1982 with an introduction by Stephen Jay Gould.

11. Gehring, W. J. (1998), *Master control genes in development and evolution: The homeobox story* (New Haven, CT: Yale University Press).

12. Morange, M. (2000), "The developmental gene concept: History and limits," in Beurton, P., Falk, R., and Rheinberger, H.-J. eds. (2000), *The concept of the gene in development and evolution* (New York: Cambridge University Press).

13. Slack, J. M. W., Holland, P. W. H., and Graham, C. F. (1993), "The zootype and the phylotypic stage," *Nature* 361: 490–492.

14. Hartwell L. H., Hopfield J. J., Leibler, S., and Murray, A. W. (1999) "From molecular to modular cell biology," *Nature* 402 (supp): C47-C52.

15. Tautz, D. (1992), "Redundancies, development and the flow of information," *BioEssays* 14: 263–266; Wilkins, A. S. (1997), "Canalization: A molecular genetic perspective," *BioEssays* 19: 257–262; Cooke, J., Nowak, M. A., Boerlijst, M., and Maynard-Smith, J. (1997), "Evolutionary origins and maintenance of redundant gene expression during metazoan development," *TIG* 13: 360–364.

16. Waddington, C. H. (1942), "Canalization of development and the inheritance of acquired characters," *Nature* 150: 563–565; Schmalhausen, I. I. (1949), *Factors of evolution: The theory of stabilizing selection* (Chicago: Chicago University Press), reprinted 1986.

17. Rudnicki, M. A., et al. (1993), "MyoD or Myf-5 is required for the formation of skeletal muscle," *Cell* 75: 1351–1359.

18. De Robertis, E. M., and Sasai, Y. (1996), "A common plan for dorsoventral patterning in Bilateria," *Nature* 380: 37–40.

19. Dickinson, W. J. (1995), "Molecules and morphology: Where's the homology?" *TIG* 11: 119–121; Bolker, J. A., and Raff, R. A. (1996), "Developmental genetics and traditional homology," *BioEssays* 18: 489–494.

20. Godwin, A. R., and Capecchi, M. R. (1998), "*Hoxc13* mutant mice lack external hair," *Genes dev.* 12: 11–20.

21. Le Mouellic, H., Lallemand, Y., and Brûlet, P. (1992), "Homeosis in the mouse induced by a null mutation in the *Hox-3.1* gene," *Cell* 69: 251–264.

22. Duboule, D. (1995), "Vertebrate *Hox* genes and proliferation: An alternative pathway to homeosis?" *Current op. gen. dev.* 5: 525–528.

23. Redline, R. W., Hudock, P., MacFee, M., and Patterson, P. (1994), "Expression of *AbdB*-type homeobox genes in human tumors," *Lab. invest.* 71: 663–670.

24. Kauffman, S. A. (1993), *The origins of order: Self organization and selection in evolution* (New York: Oxford University Press); Kauffman, S. A. (1995), *At home in the universe: The search for the laws of self-organization and complexity* (New York: Oxford University Press).

25. Gehring, W. J. (1998), *Master control genes in development and evolution: The homeobox story* (New Haven, CT: Yale University Press).

26. Quiring, R., Walldorf, U., Kloter, U., and Gehring, W. J. (1994), "Homology of the *eyeless* gene of *Drosophila* to the *small eye* gene in mice and *Aniridia* in humans," *Science* 265: 785–789.

27. Halder, G., Callaerts, P., and Gehring, W. J. (1995), "Induction of ectopic eyes by targeted expression of the *eyeless* gene in *Drosophila*," *Science* 267: 1788–1792.

28. Chow, R. L., Altmann, C. R., Lang, R. A., and Hemmati-Brivanlou, A. (1999), "Pax6 induces ectopic eyes in a vertebrate," *Development* 126: 4213–4222.

29. St-Onge, L., et al. (1997), "*Pax6* is required for differentiation of glucagon-producing α-cells in mouse pancreas," *Nature* 387: 406–409.

30. Desplan, C. (1997), "Eye development: Governed by a dictator or a junta?" *Cell* 91: 861–864; Czerny, T., et al. (1999), "*Twin of eyeless*, a second *pax-6* gene of *Drosophila*, acts upstream of *eyeless* in the control of eye development," *Molecular cell* 3: 297–307.

31. Jacob, F. (1977), "Evolution and tinkering," *Science* 196: 1161–1166; Jacob, F. (1982), *The possible and the actual* (Seattle, WA: University of Washington Press).

32. Duboule, D., and Wilkins, A. S. (1998), "The evolution of 'bricolage,'" *TIG* 14: 54–59.

33. Rose, M. R. (1991), *Evolutionary biology of aging* (New York: Oxford University Press); Klarsfeld, A., and Revah, F. (2000), *Aux origines de la mort* (Paris: Odile Jacob).

34. Lee, C.-K., Klopp, R. G., Weindruch, R., and Prolla, T. A. (1999), "Gene expression profile of aging and its retardation by caloric restriction," *Science* 285: 1390–1393; Ly, D. H., Lockhart, D. J., Lerner, R. A., and Schultz, P. G. (2000), "Mitotic misregulation and human aging," Science 287: 2486–2492.

35. Campisi, J. (1996), "Replicative senescence: An old lives' tale?" *Cell* 84: 497–500.

36. Bodnar, A. G., et al. (1998), "Extension of life-span by introduction of telomerase into normal human cells," *Science* 279: 349–352.

37. Johnson, F. B., Sinclair, D. A., and Guarente, L. (1999), "Molecular biology of aging," *Cell* 96: 291–302.

38. See earlier for the association of ApoE with Alzheimer's disease. Schächter, F. et al. (1994), "Genetic association with human longevity at the *APOE* and *ACE* loci," *Nature genetics* 6: 29–32.

39. Templeton, A. R. (1998), "The complexity of the genotype-phenotype relationship and the limitations of using genetic 'markers' at the individual level," *Science in context* 11: 373–389.

40. Yu, C.-E., et al. (1996), "Positional cloning of the Werner's syndrome gene," *Science* 272: 258–262. Shen, J.-C., and Loeb, L. A. (2000), "The Werner syndrome gene," *TIG* 16: 213–220.

41. Tower, J. (1996), "Aging mechanisms in fruit flies," *BioEssays* 18: 799–807; Lithgow, G. J. (1996), "Invertebrate gerontology: The age mutations of *Caenorhabditis elegans*," *BioEssays* 18: 809–815; Kenyon, C. (1996), "Ponce d'elegans: Genetic quest for the fountain of youth," *Cell* 84: 501–504.

42. Kimura, K. D., Tissenbaum, H. A., Liu, Y., and Ruvkun, G. (1997), "*daf-2*, an insulin receptor-like gene that regulates longevity and diapause in *Caenorhabditis elegans*," *Science* 277: 942–946.

43. Jonassen, T., et al. (1998), "Yeast Clk-1 homologue (Coq7/Cat5) is a mitochondrial protein in coenzyme Q synthesis," *J. biol. chem.* 6: 3351–3357.

44. Couzin, J. (1998), "Low-calorie diets may slow monkeys' aging," *Science* 282: 1018. Roth, G. S., Ingram, D. K., and Lane, M. A. (1999), "Calorie restriction in primates: will it work and how will we know?," *JAGS* 47: 896–903.

45. Lithgow, G. J., White, T. M., Melov, S., and Johnson, T. E. (1995), "Thermotolerance and extended life-span conferred by single-gene mutations and induced by thermal stress," *Proc. natl. acad. sci. USA* 92: 7540–7544.

46. Lin, Y.-J., Seroude, L., and Benzer, S. (1998), "Extended life-span and stress resistance in the *Drosophila* mutant *methuselah*," *Science* 282: 943–946.

47. Tower, J. (1996), "Aging mechanisms in fruit flies," *Bioessays* 18: 799–807.

48. Migliaccio, E., et al. (1999), "The p66shc adaptor protein controls oxidative stress response and life span in mammals," *Nature* 402: 309–313.

49. Martin, G. M., Austad, S. N., and Johnson, T. E. (1996), "Genetic analysis of aging: Role of oxidative damage and environmental stresses," *Nature genetics* 13: 25–34.

50. Guarente, L. (1997), "What makes us tick?" *Science* 275: 943–944.

51. Imai, S.-I., Armstrong, C. M., Kaeberlein, M., and Guarente, L. (2000), "Transcriptional silencing and longevity protein Sir2 is an NAD-dependent histone deacetylase," *Nature* 403: 795–800; Lin, S.-J., Defossez, P.-A., and Guarente, L. (2000), "Requirement of NAD and

SIR2 for life-span extension by calorie restriction in *Saccharomyces cerevisiae*," *Science* 289: 2126–2128.

52. Apfeld, J., and Kenyon, C. (1998), "Cell nonautonomy of *C. elegans daf-2* function in the regulation of diapause and life span," *Cell* 95: 199–210.

53. Hsin, H., Kenyon, C. (1999), "Signals from the reproductive system regulate the lifespan of *C. elegans*," *Nature* 399: 362–366.

54. Apfeld, J., and Kenyon, C. (1999), "Regulation of life span by sensory perception in *Caenorhabditis elegans*," *Nature* 402: 804–809.

55. Williams, G. C. (1957), "Pleiotropy, natural selection, and the evolution of senescence," *Evolution* 11: 398–411. Kirkwood, T. B. L., and Rose, M. R. (1991), "Evolution of senescence: Late survival sacrified for reproduction," *Philos. trans. roy. soc. London* ser. b *Biol. sci.* 332: 15–24.

56. Sinclair, D. A., Mills, K., and Guarente, L. (1997), "Accelerated aging and nucleolar fragmentation in yeast *sgs1* mutants," *Science* 277: 1313–1316.

57. Dawkins, R. (1995), *River out of Eden—A Darwinian view of life* (New York: Basic Books), p. 122.

58. Clarke, P. G. H., and Clarke, S. (1996), "Nineteenth century research on naturally occurring cell death and related phenomena," *Anat. embryol.* 193: 81–99.

59. Kerr, J. F. R., Wyllie, A. H., and Currie, A. R. (1972), "Apoptosis: A basic biological phenomenon with wide-ranging implications in tissue kinetics," *Br. j. cancer* 26: 239–257.

60. Raff, M. (1998), "Cell suicide for beginners," *Nature* 396: 119–122.

61. Glücksmann, A. (1951), "Cell deaths in normal vertebrate ontogeny," *Biol. rev.* 26: 59–86; Saunders, J. W., Jr. (1966), "Death in embryonic systems," *Science* 154: 604–612; Vaux, D. L., and Korsmeyer, S. J. (1999), "Cell death in development," *Cell* 96: 245–254.

62. Raff, M. C. (1992), "Social controls on cell survival and cell death," *Nature* 356: 397–400.

63. Burek, M. J., Nordeen, K. W., and Nordeen, E. J. (1995), "Initial sex differences in neuron growth and survival within an avian song nucleus develop in the absence of afferent input," *J. neurobiol.* 27: 85–96.

64. Raff, M. C., et al. (1993), "Programmed cell death and the control of cell survival: Lessons from the nervous system," *Science* 262: 695–700.

65. Thompson, C. B. (1995) "Apoptosis in the pathogenesis and treatment

of disease," *Science* 267: 1456–1462. Inhibiting cell death could be a way to fight disease: Li, M., et al. (2000), "Functional role of caspase-1 and caspase-3 in an ALS transgenic mouse model," *Science* 288: 335–339.

66. Boutibonnes, P. (1997), "La mort des bactéries: Provoquée ou programmée? Subie ou voulue?" *Médecine/sciences* 13: 73–80.

8. GENES AFFECTING BEHAVIOR

1. Dunlap, J. C. (1999), "Molecular bases for circadian clocks," *Cell* 96: 271–290.

2. Konopka, R. J., and Benzer, S. (1971), "Clock mutants of *Drosophila melanogaster*," *Proc. natl. acad. sci. USA* 68: 2112–2116; Weiner, J. (1999), *Time, love, memory: A great biologist and his quest for the origins of behavior* (New York: Alfred A. Knopf).

3. Edery, I., Rutila, J. E., and Rosbash, M. (1994), "Phase shifting of the circadian clock by induction of the *Drosophila period* protein," *Science* 263: 237–240.

4. Curtin, K. D., Huang, Z. J., and Rosbash, M. (1995), "Temporally regulated nuclear entry of the Drosophila *period* protein contributes to the circadian clock," *Neuron* 14: 365–372.

5. Sehgal, A., Price, J. L., Man, B., and Young, M. W. (1994), "Loss of circadian behavioral rhythms and *per* RNA oscillations in the *Drosophila* mutant *timeless*," *Science* 263: 1603–1606; Vosshall, L. B., et al. (1994), "Block in nuclear localization of *period* protein by a second clock mutation, *timeless*," *Science* 263: 1606–1609.

6. Sauman, I., and Reppert, S. M. (1996), "Circadian clock neurons in the silkmoth Antheraea pernyi: Novel mechanisms of period protein regulation," *Neuron* 17: 889–900.

7. Earnest, D. J., Liang, F.-Q., Ratcliff, M., and Cassone, V. M. (1999), "Immortal time: Circadian clock properties of rat suprachiasmatic cell lines," *Science* 283: 693–695.

8. Takahashi, J. S. (1994), "ICER is nicer at night (sir)," *Current biol.* 4: 165–168.

9. King, D. P., et al. (1997), "Positional cloning of the mouse circadian *clock* gene," *Cell* 89: 641–653; Antoch, M. P., et al. (1997), "Functional identification of the mouse circadian *clock* gene by transgenic BAC res-

cue," *Cell* 89: 655–667; Hogenesch, J. B., Gu, Y.-Z., Jain, S., and Bradfield, C. A. (1998), "The basic-helix-loop-helix-PAS orphan MOP3 forms transcriptionally active complexes with circadian and hypoxia factors," *Proc. natl. acad. sci. USA* 95: 5474–5479; Gekakis, N., et al. (1998), "Role of the CLOCK protein in the mammalian circadian mechanism," *Science* 280: 1564–1569.

10. Sun, Z. S., et al. (1997), "*RIGUI,* a putative mammalian ortholog of the Drosophila *period* gene," *Cell* 90: 1003–1011; Tei, H., et al. (1997), "Circadian oscillation of a mammalian homologue of the *Drosophila period* gene," *Nature* 389: 512–516; Allada, R., et al. (1998), "A mutant *Drosophila* homolog of mammalian *Clock* disrupts circadian rhythms and transcription of *period* and *timeless,*" *Cell* 93: 791–804.

11. Dunlap, J. C. (1999), "Molecular bases for circadian clocks," *Cell* 96: 271–290; Jin, X., et al. (1999), "A molecular mechanism regulating rhythmic output from the suprachiasmatic circadian clock," *Cell* 96: 57–68; Shearman, L. P., et al. (2000), "Interacting molecular loops in the mammalian circadian clock," *Science* 288: 1013–1019.

12. Ishiura, M., et al. (1998), "Expression of a gene cluster *kaiABC* as a circadian feedback process in Cyanobacteria," *Science* 281: 1519–1523.

13. Kyriacou, C. P., and Hall, J. C. (1980), "Circadian rhythm mutations in *Drosophila melanogaster* affect short-term fluctuations in the male's courtship song," *Proc. natl. acad. sci. USA* 77: 6729–6733.

14. Konopka, R. J., Kyriacou, C. P., and Hall, J. C. (1996), "Mosaic analysis in the *Drosophila* CNS of circadian and courtship-song rhythms affected by a *period* clock mutation," *J. neurogenetics* 11: 117–139.

15. Crabbe, J. C., Wahlsten, D., and Dudek, B. C. (1999), "Genetics of mouse behavior: Interactions with laboratory environment," *Science* 284: 1670–1672.

16. Cabib, S., Orsini, C., Le Moal, M., and Piazza, P. V. (2000), "Abolition and reversal of strain differences in behavioral responses to drugs of abuse after a brief experience," *Science* 289: 463–465.

17. Hall, J. C. (1994), "The mating of a fly," *Science* 264: 1702–1714; Greenspan, R. J. (1995), "Understanding the genetic construction of behavior," *Sci. Am.* 272(4): 74–79.

18. Cobb, M., and Ferveur, J.-F. (1996), "Evolution and genetic control of mate recognition and stimulation in *Drosophila,*" *Behav. proc.* 35: 35–54.

19. Ferveur, J.-F., Störtkuhl, K. F., Stocker, R. F., and Greenspan, R. J. (1995), "Genetic feminization of brain structures and changed sexual orientation in male *Drosophila*," *Science* 267: 902–905.

20. Ryner, L. C., et al. (1996), "Control of male sexual behavior and sexual orientation in Drosophila by the *fruitless* gene," *Cell* 87: 1079–1089.

21. Ferveur, J.-F., Störtkuhl, K. F., Stocker, R. F., and Greenspan, R. J. (1995), "Genetic feminization of brain structures and changed sexual orientation in male *Drosophila*," *Science* 267: 902–905.

22. LeVay, S. (1991), "A difference in hypothalamic structure between heterosexual and homosexual men," *Science* 253: 1034–1037.

23. Hamer, D. H., et al. (1993), "A linkage between DNA markers on the X chromosome and male sexual orientation," *Science* 261: 321–327.

24. Hu, S., et al. (1995), "Linkage between sexual orientation and chromosome Xq28 in males but not in females," *Nature genetics* 11: 248–256.

25. Marshall, E. (1995), "NIH's 'gay gene' study questioned," *Science* 268: 1841; Rice, G., Anderson, C., Risch, N., and Ebers, G. (1999), "Male homosexuality: Absence of linkage to microsatellite markers at Xq28," *Science* 284: 665–667.

26. Bouchard, T. J., Jr., et al. (1990), "Sources of human psychological differences: The Minnesota study of twins reared apart," *Science* 250: 223–228; Bouchard, T. J., Jr. (1994), "Genes, environment, and personality," *Science* 264: 1700–1701.

27. Balaban, E., Alper, J. S., and Kasamon, Y. L. (1996), "Mean genes and the biology of aggression: A critical review of recent animal and human research," *J. neurogenetics* 11: 1–43.

28. Ebstein, R. P., et al. (1996), "Dopamine D4 receptor (*D4DR*) exon III polymorphism associated with the human personality trait of novelty seeking," *Nature genetics* 12: 78–80; Benjamin, J., et al. (1996), "Population and familial association between the D4 dopamine receptor gene and measures of novelty seeking," *Nature genetics* 12: 81–84.

29. Zhou, Q.-Y., and Palmiter, R. D. (1995), "Dopamine-deficient mice are severely hypoactive, adipsic and aphagic," *Cell* 83: 1197–1209; Rubinstein, M., et al. (1997), "Mice lacking dopamine D4 receptors are supersensitive to ethanol, cocaine, and methamphetamine," *Cell* 90: 991–1001.

30. Lesch, K.-P., et al. (1996), "Association of anxiety-related traits with a polymorphism in the serotonin transporter gene regulatory region," *Science* 274: 1527–1531.

31. Plomin, R., and Craig, I. (1997), "Human behavioural genetics of cognitive abilities and disabilities," *BioEssays* 19: 1117–1122.

32. Lander, E. S., and Schork, N. J. (1994), "Genetic dissection of complex traits," *Science* 265: 2037–2048.

33. Risch, N. J. (2000), "Searching for genetic determinants in the new millenium," *Nature* 405: 847–856.

34. Davies, J. L., et al. (1994), "A genome-wide search for human type 1 diabetes susceptibility genes," *Nature* 371: 130–136.

35. Flint, J., et al. (1995), "A simple genetic basis for a complex psychological trait in laboratory mice," *Science* 269: 1432–1435; Melo, J. A., Shendure, J., Pociask, K., and Silver, L. M. (1996), "Identification of sex-specific quantitative trait loci controlling alcohol preference in C57BL/6 mice," *Nature genetics* 13: 147–153.

36. Skuse, D. H., et al. (1997), "Evidence from Turner's syndrome of an imprinted X-linked locus affecting cognitive function," *Nature* 387: 705–708.

37. Barker, D. (1989), "The biology of stupidity: Genetics, eugenics and mental deficiency in the inter-war years," *Brit. j. hist. sci.* 22: 347–375.

38. Herrnstein, R. J., and Murray, C. (1994), *The bell curve: Intelligence and class in American life* (New York: Free Press).

39. Gould, S. J. (1981), *The mismeasure of man* (New York: Norton).

40. Plomin, R., and DeFries, J. C. (1998), "The genetics of cognitive abilities and disabilities," *Sci. am.* 278(5): 62–69.

41. Flynn, J. R. (1999), "Searching for justice: The discovery of IQ gains over time," *Am. psychol.* 54: 5–20.

42. McClearn, G. E., et al. (1997), "Substantial genetic influence on cognitive abilities in twins 80 or more years old," *Science* 276: 1560–1563.

43. As it is true for every organism: Lewontin R. (2000), *The triple helix: Gene, organism, and environment* (Cambridge: Harvard University Press).

44. Mackintosh, N. J. (1998), *IQ and human intelligence* (Oxford: Oxford University Press).

45. Lewontin, R. C. (1976), "Race and intelligence," in Block, N. J., and Dworkin, G. eds., *The IQ controversy* (New York: Pantheon books), 78–92.

46. Devlin, B., Daniels, M., and Roeder, K. (1997), "The heritability of IQ," *Nature* 388: 468–471; Duyme, M., Dumaret, A.-C., and Tomkiewicz, S. (1999), "How can we boost IQs of 'dull children?': A late adoption study," *Proc. natl. acad. sci. USA* 96: 8790–8794.

47. "The first gene marker for IQ?" (1998) *Science* 280: 681.

48. Plomin, R., and DeFries, J. (1998), "The genetics of cognitive abilities and disabilities," *BioEssays* 19: 1117–1122.

49. Gopnik, M. (1990), "Feature-blind grammar and dysphasia," *Nature* 344: 715; Gopnik, M. (1990), "Genetic basis of grammar defect," *Nature* 347: 26; Gopnik, M., and Crago, M. B. (1991), "Familial aggregation of a developmental language disorder," *Cognition* 39: 1–50.

50. Fisher, S. E., et al. (1998), "Localisation of a gene implicated in a severe speech and language disorder," *Nature genetics* 18: 168–170.

51. Fletcher, P. (1990), "Speech and language defects," *Nature* 346: 226; Vargha-Khadem, F., and Passingham, R. E., *Nature* 346: 226; Vargha-Khadem, F., et al. (1995), "Praxic and nonverbal cognitive deficits in a large family with a genetically transmitted speech and language disorder," *Proc. natl. acad. sci. USA* 92: 930–933. Varga-Khadem, F., et al. (1998), "Neural basis of an inherited speech and language disorder," *Proc. natl. acad. sci. USA* 95: 12695–12700.

52. Cardon, L. R., et al. (1994), "Quantitative trait locus for reading disability on chromosome 6," *Science* 266: 276–279; Grigorenko, E. L., et al. (1997), "Susceptibility loci for distinct components of developmental dyslexia on chromosomes 6 and 15," *Am. j. hum. genet.* 60: 27–39.

53. In addition, there is no reason to suppose that the genetic variations that might be able to explain problems in learning to speak, read, or write are the same that explain differences between "normal" individuals. Dale, P. S., et al. (1998), "Genetic influence on language delay in two-year-old children," *Nature neuroscience* 1: 324–328.

54. Deacon, T. W. (1997), *The symbolic species: The co-evolution of language and the brain* (New York: W. W. Norton and Company). Calvin W. H., and Bickerton, D. (2000), *Lingua ex Machina: Reconciling Darwin and Chomsky with the human brain* (Cambridge: MIT Press).

55. The difficulties of defining altruism are well described in the three articles bearing on this concept in Keller, E. F., and Lloyd, E. A., eds. (1992), *Keywords in evolutionary biology* (Cambridge, MA: Harvard University Press).

56. Wilson, E. O. (1975), *Sociobiology: The new synthesis* (Cambridge, MA: The Belknap Press of Harvard University Press).

57. Hamilton, W. D. (1964), "The genetical evolution of social behavior," *J. theoret. biol.* 7: 1–52.

58. Williams, G. C. (1966), *Adaptation and natural selection* (Princeton, NJ: Princeton University Press).

59. Trivers, R. L. (1971), "The evolution of reciprocal altruism," *Quart. rev. biol.* 46: 35–57.

60. Bednekoff, P. A., and Lima, S. L. (1998), "Randomness, chaos and confusion in the study of antipredator vigilance," *TREE* 13: 284–287; Clutton-Brock, T. H., et al. (1999), "Selfish sentinels in cooperative mammals," *Science* 284: 1640–1644.

61. Holmes, W. G., and Sherman, P. W. (1982), "The ontogeny of kin recognition in two species of ground squirrels," *Amer. zool.* 22: 491–517.

62. Lewontin, R. C., Rose, R., and Kamin, L. J. (1984) *Not in our genes* (New York: Pantheon Books).

63. Waddington, C. H. (1939), *An introduction to modern genetics* (New York: Macmillan).

64. Sing, C. F., Zerba, K. E., and Reilly, S. L. (1994), "Traversing the biological complexity in the hierarchy between genome and CAD endpoints in the population at large," *Clin. genet.* 46: 6–14; Gottesman, I. I. (1997), "Twins: En route to QTLs for cognition," *Science* 276: 1522–1523.

9. Whither Genetic Determinism?

1. Gerard Toulouse kindly directed me to Anderson's article: Anderson, P. W. (1972), "More is different: Broken symmetry and the nature of the hierarchical structure of science," *Science* 177: 393–396.

2. Cline, T. W. (1989), "The affairs of *daughterless* and the promiscuity of developmental regulators," *Cell* 59: 231–234.

3. Jeffery, C. J. (1999), "Moonlighting proteins," *TIBS* 24: 8–11.

4. Hérault, Y., and Duboule, D. (1998), "Comment se construisent les doigts," *La Recherche* 305: 40–44.

5. Duboule, D., and Wilkins, A. S. (1998), "The evolution of 'bricolage,'" *TIG* 14: 54–59.

6. Sestan, N., Artavanis-Tsakonas, S., and Rakic, P. (1999), "Contact-dependent inhibition of cortical neurite growth mediated by Notch signaling," *Science* 286: 741–746.

7. Kang, S., Graham, J. M., Jr., Olney, A. H., and Biesecker, L. G. (1997), "*GLI3* frameshift mutations cause autosomal dominant Pallister-Hall syndrome," *Nature genetics* 15: 266–268.

8. Waddington, C. H. (1942), "Canalization of development and the inheritance of acquired characters," *Nature* 150: 563–565.

9. Cohen, L. G., et al. (1997), "Functional relevance of cross-modal plasticity in blind humans," *Nature* 389: 180–183.

10. Jacob, F. (1998), *Of flies, mice and men* (Cambridge, MA: Harvard University Press).

11. Alberts, B. (1998), "The cell as a collection of protein machines: Preparing the next generation of molecular biologists," *Cell* 92: 291–294; Jacob, F. (1973), *The Logic of life* (New York: Pantheon Books).

12. Jacob, F. (1982), *The possible and the actual* (Seattle, WA: University of Washington).

13. Morange, M. (1998), *A history of molecular biology* (Cambridge, MA: Harvard University Press), chapter 21.

14. Dawkins, R. (1995), *River out of Eden: A Darwinian view of life* (New York: Basic books), 109.

15. Kauffman, S. A. (1993), *The origins of order: Self organization and selection in evolution* (New York: Oxford University Press); Kauffman, S. A. (1995), *At home in the universe: The search for the laws of self-organization and complexity* (New York: Oxford University Press).

16. Bock, G. R., and Goode, J. A., eds (1998), *The limits of reductionism in biology*, Novartis Foundation Symposium 213 (Chichester: John Wiley and sons).

17. Uetz, P., et al. (2000), "A comprehensive analysis of protein-protein interactions in *Saccharomyces cerevisiae*," *Nature* 403: 623–627.

Eisenberg, D., Marcotte, E. M., Xenarios, I., and Yeates, T. O. (2000), "Protein function in the post-genomic era," *Nature* 405: 823–826.

18. Brown, P. O., and Botstein, D. (1999), "Exploring the new world of the genome with DNA microarrays," *Nature genetics supplement* 21: 33–37.

19. Thomas, S. A., and Palmiter, R. D. (1997), "Impaired maternal behavior in mice lacking norepinephrine and epinephrine," *Cell* 91: 583–592.

20. Muller, H. J. (1926), "The gene as the basis for life," in Muller, H. J. (1962), *Studies in genetics: The selected papers of H. J. Muller,* (Bloomington, IN: Indiana University Press), 196.

21. Deamer, D. W., and Fleischaker, G. R. (1994), *Origins of life: The central concepts* (Boston: John and Bartlett Publishers); Orgel, L. E. (1998), "The origin of life—a review of facts and speculations," *TIBS* 23: 491–495.

10. HUMAN EVOLUTION AND EUGENICS

1. Muller-Hill, B. (1997), *Murderous science: Elimination by scientific selection of Jews, Gypsies and others in Germany, 1933–1945* (Cold Spring Harbor, NY: Cold Spring Harbor Laboratory Press).

2. Cavalli-Sforza, L. L., Menozzi, P., and Piazza, A. (1994), *The history and geography of human genes* (Princeton, NJ: Princeton University Press); Marks, J. (1995), *Human biodiversity: Genes, race and history* (New York: Aldine de Gruyter).

3. Erlich, H. A., Bergström, T. F., Stoneking, M., and Gyllensten, U. (1996), "HLA sequence polymorphism and the origin of humans," *Science* 274: 1552–1554.

4. Krings, M., et al. (1997), "Neandertal DNA sequences and the origin of modern humans," *Cell* 90: 19–30. Ovchinnikov, I. V., et al. (2000), "Molecular analysis of Neanderthal DNA from the northern Caucasus," *Nature* 404: 490–493.

5. Cavalli-Sforza, L. L. (2000), *Genes, peoples, and languages* (New York: North Point Press (Farrar, Straus and Giroux)).

6. Semino, O., et al. (2000), "The genetic legacy of paleolithic *Homo sapiens sapiens* in extant Europeans: a Y chromosome perspective," *Science* 290: 1155–1159.

7. Feldman, M. W., and Laland, K. N. (1996), "Gene-culture coevolutionary theory," *TREE* 11: 453–457; Aoki, K. (1986), "A stochastic model of gene-culture coevolution by the 'culture historical hypothesis' for the evolution of adult lactose absorption in humans," *Proc. natl. acad. sci. USA* 83: 2929–2933.

8. Kevles, D. (1985), *In the name of eugenics* (New York: Alfred A. Knopf).

9. Duster, T. R. (1990), *Backdoor to eugenics* (New York: Routledge, Chapman and Hall Inc.).

10. McLaren, A. (2000), "Cloning: pathways to a pluripotent future," *Science* 288: 1775–1780.

11. Illmensee, K., and Hoppe, P. C. (1981), "Nuclear transplantation in mus musculus: Developmental potential of nuclei from preimplantation embryos," *Cell* 23: 9–18.

12. McGrath, J., and Solter, D. (1984), "Inability of mouse blastomere nuclei transferred to enucleated zygotes to support development in vitro," *Science* 226: 1317–1319.

13. Sims, M., and First, N. L. (1993), "Production of calves by transfer of nuclei from cultured inner cell mass cells," *Proc. natl. acad. sci. USA* 90: 6143–6147.

14. Campbell, K. H. S., McWhir, J., Ritchie, W. A., and Wilmut, I. (1996), "Sheep cloned by nuclear transfer from a cultured cell line," *Nature* 380: 64–66.

15. Wilmut, I., et al. (1997), "Viable offspring derived from fetal and adult mammalian cells," *Nature* 385: 810–813.

16. In fact, cloning is a "node" of biological, social, and ethical issues.

17. For a discussion of the numerous possibilities of use of the embryonic stem cells, see the collection of articles gathered in: (2000) "Stem cells branch out," *Science* 287: 1417–1446.

18. It was, in fact, the main objective of their authors: Wilmut, I., Campbell, K., and Tudge, C. (2000), *The second creation: the age of biological control by the scientists who cloned Dolly* (London: Headline).

19. Schnieke, A. E., et al. (1997), "Human factor IX transgenic sheep produced by transfer of nuclei from transfected fetal fibroblasts," *Science* 278: 2130–2133.

20. Duster, T. R. (1990), *Backdoor to eugenics* (New York: Routledge, Chapman and Hall Inc.).

21. Pier, G. B., et al. (1998), "*Salmonella typhi* uses CFTR to enter intestinal epithelial cells," *Nature* 393: 79–82.

22. Jacob, F. (1999), *Of flies, mice and men* (Cambridge MA: Harvard University Press), p. 188.

23. Strohman, R. C. (1999), "Five stages of the Human Genome Project," *Nature biotechnology* 17: 112.

Index